High Performance with Laravel Octane

Learn to fine-tune and optimize PHP and Laravel apps using Octane and an asynchronous approach

Roberto Butti

BIRMINGHAM—MUMBAI

High Performance with Laravel Octane

Copyright © 2022 Packt Publishing

Group Product Manager: Pavan Ramchandani
Publishing Product Manager: Bhavya Rao
Senior Editor: Mark D'Souza
Senior Content Development Editor: Debolina Acharyya
Technical Editor: Saurabh Kadave
Copy Editor: Safis Editing
Project Coordinator: Sonam Pandey
Proofreader: Safis Editing
Indexer: Tejal Daruwale Soni
Production Designer: Prashant Ghare
Marketing Coordinator: Anamika Singh

First published: December 2022
Production reference: 1081222

Published by Packt Publishing Ltd.
Livery Place
35 Livery Street
Birmingham
B3 2PB, UK.

ISBN 978-1-80181-940-4

www.packt.com

To my family, for supporting and inspiring me.

– Roberto Butti

Contributors

About the author

Roberto Butti has been working in web development since 2000. He has worked for different markets and companies using several languages/frameworks in both server- and client-side web development.

Today, he is focused on web architecture with Laravel on the server side and various reactive Javascript frameworks, such as Vue and Svelte, on the client side.

Always passionate about performant and scalable architectures, he has found – in Laravel Octane and, more generally, in Open Swoole – an enabling tool for projects where concurrency, real-time communication, and performance are crucial.

About the reviewers

Fabio Ivona is an Italian software engineer, proud husband, and father of three sweet children, working as a full-stack developer with PHP and JS since 2005. He came across Laravel (*v4*) in 2013 and immediately fell in love with it, exploring all its cool features every day.

His workstation is an Ubuntu Linux laptop, and his everyday toolbox consists of Laravel, Livewire/Inertia, Tailwind, Alpine.js, Vite, Angular, and React, as well as Docker and Kubernetes.

He's currently the founder/CTO at def.studio, a web development company. He's a member of the Pest PHP core team and regularly contributes to open source projects on his own and fellow developers' repositories.

Andrea Marco Sartori is an Italian software engineer who lives in Australia. He has over 16 years of experience as a web developer and over 9 years of experience with the Laravel framework.

He has worked on a number of projects for several international organizations, from start-ups to large companies, including Yahoo!

He is an active member of the Laravel community and likes to contribute to the Laravel ecosystem by sending pull requests to the core or by creating free packages that are downloaded hundreds of thousands of times.

Table of Contents

3

Part 3: Laravel Octane – a Complete Tour

4

5

Reducing Latency and Managing Data with an Asynchronous Approach   115

Part 4: Speeding Up

6

Using Queues to Apply the Asynchronous Approach in Your Application   133

7

Configuring the Laravel Octane Application for the Production Environment   155

Index   177

Other Books You May Enjoy   184

Preface

This book provides a 360-degree overview of what it takes to design and build a performant application with Laravel Octane, why you should, and how to do it. This book also covers the different tools used by Laravel Octane, such as Swoole and RoadRunner, and lists and explains the various features and differentiating elements. But the most important thing is that it will enable you to understand why and when to use Swoole or RoadRunner.

Who this book is for

This book is for those Laravel developers who already know the fundamentals of the framework, such as routing mechanisms, controllers, services, and models. It is meant for those developers who would like to design their applications in a more scalable way and make them more performant.

What this book covers

Chapter 1, Understanding the Laravel Web Application Architecture, is where Laravel Octane introduces new features, but more importantly a new way of thinking about how to design high-performance Laravel applications.

Chapter 2, Configuring the RoadRunner Application Server, shows us a scenario where Laravel Octane is based on the RoadRunner application server. RoadRunner is a very fast and effective application server that is easy to install. Because of its simplicity, it allows the user to approach Laravel Octane easily and straightforwardly.

Chapter 3, Configuring the Swoole Application Server, shows us a scenario where Laravel Octane is based on the Swoole application server. Running Swoole as an application server allows the developer to use some advanced features, such as managing multiple workers and concurrent tasks, bootstrapping the application more efficiently, fast caching, and sharing data among workers. Laravel Octane uses Swoole, so it is important to have clear ideas about the opportunities that arise from using these application servers.

Chapter 4, Building a Laravel Octane Application, teaches us how to build an application with practical examples using features provided by the Swoole application server.

Chapter 5, *Reducing Latency and Managing Data with an Asynchronous Approach*, teaches us that, once the developer has sped up the bootstrapping of the framework and has removed bottlenecks on the request side, they have to reduce latency in other parts of the application. The chapter introduces useful techniques to reduce latency in executing tasks and managing data.

Chapter 6, *Using Queues to Apply the Asynchronous Approach in Your Application*, teaches us that in addition to the various tools available with application servers, we can add extra components to our application architecture. One of the main tools to go alongside application servers for greater scalability of a solution is queuing mechanisms. This chapter shows what the benefits are of adding a queuing mechanism to our application.

Chapter 7, *Configuring the Laravel Octane Application for the Production Environment*, discusses going to production and covers environment-specific configurations, deploying an application to an application server, and fine-tuning the configuration of nginx as a reverse proxy for Swoole.

To get the most out of this book

You will need PHP 8.0 or above installed on your computer. All code examples have been tested using PHP 8.1 on macOS and GNU/Linux.

Software/hardware covered in the book	Operating system requirements
Laravel 9	Windows, macOS, or Linux
PHP 8+	

If you don't want to have PHP installed on your local machine and you are familiar with the Docker setup, you could use the Docker images provided by Laravel Sail. The book explains both local installation and installation via Docker images for the setup of the application servers and the modules (Swoole/OpenSwoole).

If you are using the digital version of this book, we advise you to type the code yourself or access the code from the book's GitHub repository (a link is available in the next section). Doing so will help you avoid any potential errors related to the copying and pasting of code.

Download the example code files

You can download the example code files for this book from GitHub at `https://github.com/PacktPublishing/High-Performance-with-Laravel-Octane`. If there's an update to the code, it will be updated in the GitHub repository.

Download the color images

We also provide a PDF file that has color images of the screenshots and diagrams used in this book. You can download it here: `https://packt.link/ZTNyn`

Conventions used

There are a number of text conventions used throughout this book.

`Code in text`: Indicates code words in text, database table names, folder names, filenames, file extensions, pathnames, dummy URLs, user input, and Twitter handles. Here is an example: "From the Symfony world, Laravel includes packages such as `Symfony/routing` to manage routing, and `http-foundation` and `http-kernel` to manage HTTP communication."

A block of code is set as follows:

```
"nyholm/psr7": "^1.5",
"spiral/roadrunner": "v2.0"
```

When we wish to draw your attention to a particular part of a code block, the relevant lines or items are set in bold:

```
command=/usr/bin/php -d variables_order=EGPCS /var/www/html/
artisan octane:start --server=swoole --host=0.0.0.0 --port=80
--watch
```

Any command-line input or output is written as follows:

```
npm install --save-dev chokidar
```

Bold: Indicates a new term, an important word, or words that you see onscreen. For instance, words in menus or dialog boxes appear in **bold**. Here is an example: "Once everything is fine, you will see **Table created!** on your web page. This means that the rows were created in the right way."

> **Tips or important notes**
> Appear like this.

Get in touch

Feedback from our readers is always welcome.

General feedback: If you have questions about any aspect of this book, email us at `customercare@packtpub.com` and mention the book title in the subject of your message.

Errata: Although we have taken every care to ensure the accuracy of our content, mistakes do happen. If you have found a mistake in this book, we would be grateful if you would report this to us. Please visit `www.packtpub.com/support/errata` and fill in the form.

Piracy: If you come across any illegal copies of our works in any form on the internet, we would be grateful if you would provide us with the location address or website name. Please contact us at copyright@packt.com with a link to the material.

If you are interested in becoming an author: If there is a topic that you have expertise in and you are interested in either writing or contributing to a book, please visit authors.packtpub.com.

Share Your Thoughts

Once you've read, we'd love to hear your thoughts! Scan the QR code below to go straight to the Amazon review page for this book and share your feedback.

https://packt.link/r/1801819408

Your review is important to us and the tech community and will help us make sure we're delivering excellent quality content.

Download a free PDF copy of this book

Thanks for purchasing this book!

Do you like to read on the go but are unable to carry your print books everywhere? Is your eBook purchase not compatible with the device of your choice?

Don't worry, now with every Packt book you get a DRM-free PDF version of that book at no cost.

Read anywhere, any place, on any device. Search, copy, and paste code from your favorite technical books directly into your application.

The perks don't stop there, you can get exclusive access to discounts, newsletters, and great free content in your inbox daily

Follow these simple steps to get the benefits:

1. Scan the QR code or visit the link below

https://packt.link/free-ebook/9781801819404

2. Submit your proof of purchase
3. That's it! We'll send your free PDF and other benefits to your email directly

Part 1:
The Architecture

The objective of this part is to familiarize you with the application server architecture for PHP and show how the application server shares resources across the workers to serve HTTP requests.

This part comprises the following chapters:

- *Chapter 1, Understanding the Laravel Web Application Architecture*

1
Understanding the Laravel Web Application Architecture

This book is for **Laravel** developers who would like to make or design their Laravel web application in a more scalable way and make it more performant.

This book aims to provide you with knowledge, suggestions, and explanations on how to improve the software architecture of a web application, starting from a typical PHP web application architecture to a more scalable and performant architecture.

It provides a 360-degree overview of what it takes to design and build a performant application with **Laravel Octane**. We'll see why Laravel Octane is suitable for designing and building a performant application. The book also covers the different tools used by Laravel Octane, such as **Open Swoole** and **RoadRunner**, and lists and describes the various features and differentiating elements. But the most important thing is to enable you to understand why and when to use Open Swoole or RoadRunner.

But before starting, why use Laravel Octane?

Laravel Octane is a tool that allows us to access some functionality and features exposed by the two application servers we just mentioned.

One of the benefits of Laravel Octane is the huge improvement in the response times to HTTP requests by clients such as web browsers. When we develop a Laravel application, we use an important software layer implemented by the framework. This software layer needs time and resources to start. Even if we talk about only a few resources and a short amount of time, this repeated action for each request, especially in a context where there are many requests, can be a problem. Or rather, its optimization could bring enormous benefits.

Laravel Octane, through application servers, does just that: optimizes the process of starting the framework, which typically happens for each individual request. We will see in detail how this is done; essentially, the objects and everything that the framework needs are started and allocated to the start of the application server, and then the instances are made available to the various workers. **Workers** are the processes that are initiated to serve the requests.

Another reason why it is interesting to evaluate the adoption of Laravel Octane in your Laravel web application is that by using an application server such as Swoole, you can access those features implemented by Swoole.

The functions are, for example, the advanced mechanisms of the cache driver, shared storage for sharing information between the various workers, and the execution of tasks in parallel.

This is a totally new concept to the classic PHP developer who typically does not have an immediate functionality available in the PHP core for the parallelization of processes.

This chapter will introduce you to the Laravel ecosystem and explore what Laravel Octane is.

This chapter aims to introduce you to the application server approach, in which more workers cooperate to manage multiple requests. Understanding the behavior under the hood allows the developer to avoid some mistakes, especially on shared resources (objects and global states) across the worker. This is important because the classical PHP approach is to have one dedicated thread to manage one request.

In this chapter, we will cover the following topics:

- Exploring the Laravel ecosystem
- Understanding the request lifecycle
- Getting to know the application server

Technical requirements

In order to run the code and tools shown in this book, you must have the **PHP engine** installed on your machine. It is recommended that you have a recent version of PHP installed (at least *8.0*, released in November 2020).

Also, in order to easily install additional tools, it is recommended that you have Homebrew installed if you use macOS. In the case of GNU/Linux systems, it will be sufficient to resort to using the package manager of the distribution used, and in the case of Windows systems, the advice is to have a virtual environment, for example, with Docker.

In the current chapter, some commands and source code will be shown, simply to share some concepts. In subsequent chapters, especially the second chapter about RoadRunner and the third chapter about Open Swoole, the installation of each package and tool will be addressed step by step.

There are those who, regardless of their operating system, prefer to maintain a "clean" installation by resorting to using Docker regardless of the host operating system. Where the next chapters deal with the installation of operating system-dependent tools, the different methods will be highlighted depending on the system in use.

The source code and the configuration files of the examples described in the current chapter are available here: https://github.com/PacktPublishing/High-Performance-with-Laravel-Octane/tree/main/octane-ch01

Exploring the Laravel ecosystem

Laravel is a great framework in the PHP ecosystem that helps developers to build web applications quickly and reliably.

It includes, as dependencies, some great tools from the PHP ecosystem, such as **Symfony** packages, and some other solid and mature packages such as **Monolog** for logging, **Flysystem** for accessing files and storage, and **CommonMark** for managing Markdown format.

From the Symfony world, Laravel includes packages such as `Symfony/routing` to manage routing, and `http-foundation` and `http-kernel` to manage HTTP communication.

All this is just to say that Laravel uses the best parts of the PHP ecosystem, puts them together, and provides tools, helpers, classes, and methods to simplify the usage of all the tools from the developer's perspective.

In addition, Laravel is more than a framework. Laravel is an ecosystem.

Laravel also provides applications and services that are integrated with the framework.

For example, Laravel provides the following:

- **Cashier**: For integration with **Stripe** and **Paddle** for the payment and subscription process.
- **Breeze**, **Jetstream**, **Sanctum**, and **Socialite**: For managing authorization, authentication, the social login integration process, and exposing protected APIs.
- **Dusk** and **Pest**: For testing.
- **Echo**: For broadcasting events in real time.
- **Envoyer**, **Forge**, and **Vapor**: For server or serverless management and to manage the deployment process.
- **Mix**: For compiling **JavaScript** and **CSS** through a webpack configuration fully integrated with the Laravel frontend.
- **Horizon**: A web user interface for monitoring queues based on **Redis**.
- **Nova**: An administrator panel builder for Laravel applications.
- **Sail**: A local development environment based on **Docker**.
- **Scout**: A full-text search engine, backed by providers such as **Algolia**, **Meilisearch**, or simply by the **MySQL** or **PostgreSQL** database.
- **Spark**: A boilerplate solution for managing billing/subscription in your application.
- **Telescope**: UI module for showing debugging and insights.

- **Valet**: A macOS-specific bundle of applications configured for running the PHP application. It has dependencies with **nginx**, **PHP**, and **Dnsmasq**.

- **Octane**: For improving performance and optimizing resources.

In this book, we will analyze the last tool in this list: Laravel Octane.

We will go over the use of other tools within the Laravel ecosystem, such as Sail (for simplifying the installation process of a complete development environment), and Valet (for correctly setting up the local environment to run a web server and PHP). Also, Laravel Octane depends on important software that we will see in-depth throughout the book. Laravel Octane has strong requirements: it requires additional software such as Swoole or RoadRunner.

But one step at a time.

Before we delve into the tools and their configuration, it's important to understand some basic mechanisms for managing **HTTP requests**.

HTTP

HTTP is a protocol that defines rules, messages, and methods for fetching resources on the web, such as HTML documents (web pages) and assets. Clients (who require the resource) and servers (who serve the resource) communicate by exchanging messages. The client sends requests, and the server sends responses.

One of the goals of the book is to empower you to improve the performance of your web applications by doing different things, starting with designing the architecture of the application, choosing and using the right tools, writing code, and finally, releasing the application.

The tools we are going to analyze and use will do much of the work, but I think it is important to understand the underlying dynamics to have a good awareness of how the various tools work to enable you to configure, integrate, and use them to the best of their ability.

Before we get deeper into the workings of Laravel Octane, let me take you through how servers typically deal with HTTP requests by explaining the HTTP request lifecycle.

Understanding the HTTP request lifecycle

There are a number of components involved in performing the HTTP request. The components are as follows:

- **Client**: This is where the request starts and the response ends (for example, the browser).

- **Network**: The requests and responses go through this and it connects the server and the client.

- **Proxy**: This is an optional component that could perform some tasks before the request reaches the web server, such as caching, rewriting and/or altering the request, and forwarding the request to the right web server.

- **Web server**: This receives the request and is responsible for selecting the correct resource.

- **PHP**: The language, or more generally in the case of server-side languages, the language-specific engine that is used and involved. In this case, the PHP interpreter is used. The PHP interpreter can be activated mainly in two ways: as a web server module or as a separate process. In the latter case, a technology called **FastCGI Process Manager** (**FPM**) is used. We will see how this mechanism works later in more detail. For now, it is useful to know that the web server somehow invokes the server-side language interpreter. By doing this, our server is able to interpret the language. If the invoked resource is a PHP-type file with the specific PHP syntax, the resource file requested is interpreted by the PHP engine. The output is sent back in the form of a response to the web server, the network, and then the browser.

- **Framework**: In the case that the application is written with PHP and a framework is used, as a developer, you can access classes, methods, and helpers to build your application faster.

The components are called sequentially in the HTTP request flow. The HTTP request starts from the browser, then goes through the network (optionally passing through via the proxy), until it reaches the web server that invokes the PHP engine and the framework is bootstrapped.

From a performance point of view, if you want to bring some improvement, you have to take some action or implement some solution depending on the elements of this architecture.

For example, on the browser side, you could work on caching assets in the browser or on the optimization of your JavaScript code. On the networking side, one solution could be resource optimization, for example, reducing the weight of assets or introducing architectural elements such as a CDN. In the case of the web server, an effective first-level improvement could be to avoid loading the PHP engine for the static assets (non-PHP files).

All such fine-tuning will be addressed in the final chapters, where we will deal with the configuration and optimization of production elements. Most of the book covers the optimization of the framework. For example, in *Chapters 2* and *3*, topics such as the use of Octane with tools such as Swoole and RoadRunner, which enable more efficient and effective loading of resources (shared objects and structures), are addressed. Other points of performance improvement on the framework side include the introduction of an asynchronous approach through the use of queuing systems (*Chapters 6* and *7*).

Now that you have an idea of the components involved in an HTTP request, let's look at the structure of an HTTP request.

The structure of an HTTP request

To understand in detail what happens in a typical HTTP request, we start by analyzing what is sent from the browser to the server during a request. A request is mainly characterized by methods (GET, POST, PUT, and so on), the URL, and HTTP headers.

The URL is visible in the browser's address bar, whereas the headers are handled automatically by the browser and are not directly visible to the user.

The following describes the structure of an HTTP request:

- The **HTTP method** (or **HTTP verbs**) in an HTTP request represents the operation the frontend side wants to perform on the server side with the requested resource:

  - The GET method: Reads and retrieves a resource

  - The POST method: Creates a new resource

  - The PUT method: Replaces a resource

  - The PATCH method: Edits the resource

  - The DELETE method: Deletes the resource

- The **URL** identifies the resource. We'll explain the URL structure in the next section (*Handling an HTTP request*).

- The **headers** include additional information that allows the server to understand how to handle the resource. This information can comprise authentication information, the required format of the resource, and so on.

- The **body payload** is additional data, for example, the data sent when a form is submitted to the server.

Now that you have an idea of the structure of an HTTP request, let's see how such requests are handled.

Handling an HTTP request

A URL is made up of the protocol, the hostname, the port, the path, and the parameters. A typical URL is as follows:

```
<protocol>://<hostname>:<port>/<path>?<parameters>
```

For example, a URL could be the following:

```
https://127.0.0.1:8000/home?cache=12345
```

Each part that makes up the HTTP request is used specifically by the various software involved in handling the HTTP request:

- A protocol is used by the browser to determine the communication encryption (encrypted via HTTPS or non-encrypted via HTTP).

- A hostname is used by the DNS to resolve the hostname into an IP address, and by the web server to involve the right virtual host.

- A port is used by the operating system of the server to access the right process.

- A path is used by the web server to call the right resource and for the framework to activate the routing mechanism.

- Parameters are used by the application to control the behavior of the logic (server-side for query parameters and client-side for the anchor parameters). For example, the query parameters are defined after the ? character, and the anchor parameters are defined after the # character in the URL: `https://127.0.0.1:8000/?queryparam=1#anchorparam`.

First, the protocol (typically HTTP or HTTPS) is defined. Next, the hostname, which is useful for figuring out which server to contact, is specified. Then, there is a part that is not normally specified, which is the port number; typically, it is port 80 for HTTP and 443 for HTTPS. Also present is the path that identifies the resource we are requesting from the server. Finally, two other optional parts deal with parameters. The first concerns server-side parameters (query string), and the second concerns client-side or JavaScript parameters (parameters with anchors).

In addition to the URL, another characteristic element of the request is the HTTP header, which is very important for the server reached by the request to better understand how to handle the request.

HTTP headers are automatically handled by the browser based on some information and browsing state. Typically, the headers specify the format and other information about the resource; for example, they specify the MIME type, user agent, and so on.

They also specify any access tokens in case the requested resource is protected. The elements to manage the state are also present in the headers as cookies and references for the session. This information is useful for the server to understand and relate consecutive requests.

Why is it so important to understand how a request is composed? Because in analyzing optimization elements regarding performance, the structure of the URL and the parts that make up the headers determine the behavior of different elements within the web architecture (browser, network, web server, server-side language, and framework).

For example, an element such as a hostname is useful to the DNS (network) to be able to resolve the hostname into the IP address. Knowing this is useful in deciding whether to do caching, for example, for name resolution.

Each element involved has its own characteristics that can be optimized to be able to get better performance.

One of the characterizing elements of a typical request to a classic PHP application is that each request is independent of any other request. This means that if your PHP script instantiates an object, this operation is repeated with each request. This has little impact if your script is called only a few times and your script is simple.

Let's try to think of a scenario in which we have a framework-based application, with the application having to deal with a high load of concurrent requests.

A framework-based application has numerous objects at its disposal, which must be instantiated and configured at startup. In the classic case of PHP, the startup of the framework corresponds to a request.

Laravel Octane, on the other hand, introduces a new way of starting the application.

In a classic Laravel web application, it is sufficient to have a web server (such as nginx) or the internal web server provided by Laravel in the case of development on the developer's local computer. A classic web server can handle requests without any kind of resource-sharing unless these resources are external resources such as a database or a cache manager.

In contrast to what happens with a classic web server, an application server has the task of starting and managing the executions of multiple workers. Each worker will be able to handle multiple requests by reusing objects and parts of the logic of your application.

This has one benefit, which is that the actual startup of your application and the setting up of the various objects occur on the first request received from the worker and not on each individual request.

HTTP requests and Laravel

From the Laravel application perspective, the parts involved directly in the HTTPS requests are typically routes and controllers.

Handling a request through a Laravel application typically means having to implement the routing part and implement the logic to manage the request in the controller. Routing allows us to define which code to execute within our Laravel application against a specific path in the URL. For example, we might want to define that the code of a method in a specific class such as `HomeController::home()` must be invoked against a request that has a `/home` path in the URL. In the classic Laravel definition, we would write something like this in the `routes/web.php` file:

```
Route::get('/home', [HomeController::class, 'home'])-
>name("home");
```

Now we have to implement the logic to manage the request in the `HomeController` class (that we have to create) and implement the `home` method. So, in a new `app/Http/Controllers/HomeController.php` file, you have to implement the `HomeController` class extending the basic controller class:

```php
<?php

namespace App\Http\Controllers;

class HomeController extends Controller
{
    public function home(): string
    {
        return "this is the Home page";
    }
}
```

Now that you have an understanding of how web servers handle requests, let us learn more about the application servers that Laravel Octane integrates with.

Getting to know the application server for Laravel Octane

In the PHP ecosystem, we have several application servers.

Laravel Octane, which handles server configuration, startup, and execution, integrates mainly with two of them: Swoole and RoadRunner.

We will deal with the installation, configuration, and use of these two application servers in detail later on.

For now, it is enough for us to know that once the application servers are installed, Laravel Octane will take care of their management. Laravel Octane will also take care of their proper startup via the following command:

```
php artisan octane:start
```

The `octane:serve` command is added when Laravel Octane is installed.

In other words, Laravel Octane has a strong dependency on application servers such as RoadRunner or Swoole.

At startup, Laravel Octane via Swoole or RoadRunner activates some workers, as shown in the following figure:

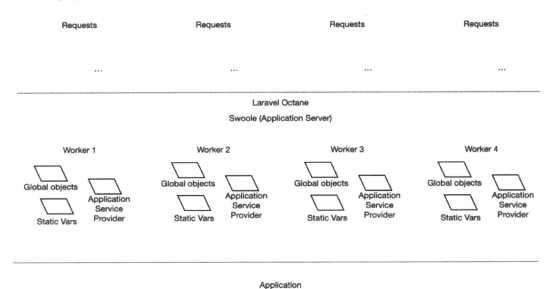

Figure 1.1: The activation of workers

What are workers?

In Octane, a **worker** is a process that takes charge of handling the requests associated with it. A worker has the responsibility of starting the framework and initializing framework objects.

This has an extremely positive impact from a performance standpoint. The framework is instantiated on the first request assigned to the worker. The second (and subsequent) requests assigned to that worker reuse the objects already instantiated. The side effect of this is that the worker shares instances of global objects and static variables between requests. This means that different calls to the controller can access the data structures that are shared between requests.

To complicate matters, there is the fact that requests assigned to the same worker share a global state, but different workers are independent and have scope independent of each other. So, we can say that not all requests share the same global state. Requests share a global state when associated with the same worker. Two requests from two different workers share nothing.

In order to minimize the side effect, Laravel Octane has the responsibility of managing the reset of classes/objects owned directly by the framework across the requests.

However, Octane can't manage and reset classes owned directly by the application.

That's why the main thing to pay attention to when using Octane is the scope and lifecycle of variables and objects.

To understand this better, I will give you a very basic example.

Example with a shared variable

This example, in the routes/web.php file, creates a route for path / and returns a human-readable timestamp. To simplify the explanation, we are going to write the logic directly into the route file instead of calling and delegating the logic to a controller:

```
$myStartTime = microtime(true);
Route::get('/', function () use($myStartTime) {
    return DateTime::createFromFormat('U.u', $myStartTime)
    ->format("r (u)");
});
```

In the routes/web.php routing file (web.php is already stored in the routes directory in the Laravel root folder project), a $myStartTime variable is instantiated and assigned the current time expressed in milliseconds. This variable is then inherited by the route/management function via the use clause.

In the performance of the function associated with route/, the contents of the $myStartTime variable are returned and then displayed.

With the classic behavior of the Laravel application, at each invocation/execution, the variable is regenerated and reinitialized (each time with a new value).

To start the Laravel application in the classic mode, simply run the following:

```
php artisan serve
```

Once the web server is started, go to the following URL via the browser: http://127.0.0.1:8000

By continuously reloading the page, a different value is displayed each time; basically, the timestamp is displayed with each request.

Instead of using the development web server provided by Laravel, you would use Laravel Octane and have a different result. At each page refresh (reloading of the web page), you would always see the same value. The value is relative to the timestamp of the first request served. This means that the variable is initialized with the first request and then the value is reused across the requests.

If you try to refresh multiple times, in some cases, you could see a new value.

If this happens, it means that the request was managed by the second (or a new) worker. This means that this behavior is quite unpredictable because Octane acts as a load balancer. When a request comes from the network, the application server will decide which worker (of those available) to assign the request to.

In addition to this, another element that could cause a new value to be generated is when you hit the maximum number of requests managed by a single worker. We will see how to define the maximum number of requests later, and in general, we will have a deep dive session (in *Chapters 2* and *3*) into Laravel Octane configuration.

The behavior whereby variables are shared across workers until the application server is restarted is valid only for global variables or objects stored in the application service container. The local variables (the variables for which the scope is limited to a function or a method) are not affected.

For example, in the code previously shown, I'm going to declare a $myLocalStartTime variable in the function called by the routing mechanism. The scope of the $myLocalStartTime variable and its lifecycle is limited to the Closure function:

```
$myStartTime = microtime(true);
Route::get('/', function () use($myStartTime) {
    $myLocalStartTime = microtime(true);
    return DateTime::createFromFormat('U.u', $myStartTime)
    ->format("r (u)") . " - " .
    DateTime::createFromFormat('U.u', $myLocalStartTime)
    ->format("r (u)");
});
```

Execute the following command with the classic Laravel web server:

```
php artisan serve
```

You will see that both values will change on each new request. You can see that when you open a browser to http://127.0.0.1:8000.

Launch Octane as a server with the following command:

```
php artisan octane:start
```

You will see, in your browser at http://127.0.0.1:8000, two different dates/times with milliseconds. If you refresh the page, you will see a change in just the second one ($myLocalStartTime).

You have to be aware of this behavior when you are building an application based on Octane.

Another example to better understand this behavior is creating a class with a static property.

Creating a class with a static property

In order to keep this example as simple as possible, I created a MyClass class in the routes/web.php file.

I'm going to add new routes that call the add() method of the MyClass object and then call and return the value of the static property retrieved by the get() method.

In routes/web.php, add the following class:

```
class MyClass
{
    public static $number = 0;

    public function __construct()
    {
        print "Construct\n";
    }
    public function __destruct()
    {
        print "Deconstruct\n";
    }
    public function add()
    {
        self::$number++;
    }
    public function get()
    {
        return self::$number;
    }
}
```

Then, in the routes/web.php file, declare the new route as follows:

```
Route::get('/static-class', function (MyClass $myclass) {
    $myclass->add();
    return $myclass->get();
});
```

Next, you can launch Laravel in a classic way using the following command:

```
php artisan serve
```

Now, if you access the URL http://127.0.0.1:8000/static-class multiple times, the value 1 will be shown. This is because, classically, for every request, the MyClass object is instanced from scratch.

Launch Laravel Octane using the following command:

```
php artisan octane:serve
```

If you then access the URL http://127.0.0.1:8000/static-class multiple times, you will see the value 1 in the first request, 2 in the second, 3 in the third, and so on. This is because, with Octane, MyClass is instanced for every request, but the static values are kept in memory.

With a non-static property, we can see the difference as follows:

```
class MyClass
{
    public static $numberStatic = 0;
    public $number = 0;
    public function __construct()
    {
        print "Construct\n";
    }
    public function __destruct()
    {
        print "Deconstruct\n";
    }
    public function add()
    {
        self::$numberStatic++;
        $this->number++;
    }
    public function get()
    {
        return self::$numberStatic . " - " . $this->number;
    }
}
```

After calling the page five times, the result shown in the browser will be as follows:

Construct Deconstruct 5 – 1

This is quite simple but, in the end, good for understanding the behavior of static variables under the hood.

The use of static variables is not so unusual. Just think of singleton objects or the app container of Laravel.

To avoid unexpected behavior – as in this specific example with static variables but more generally, with global objects (Laravel makes extensive use of them) – explicit re-initialization must be taken care of. In this case, the static variable is initialized in the constructor. My suggestion is to use explicit initialization of the properties in the constructor. This is because it is the developer's responsibility to take care of the re-initialization of the variables in the case of global states (objects and variables).

```
class MyClass
{
    public static $numberStatic = 0;
    public $number = 0;
    public function __construct()
    {
        print "Construct\n";
        self::$numberStatic = 0;
    }
    public function __destruct()
    {
        print "Deconstruct\n";
    }
    public function add()
    {
        self::$numberStatic++;
        $this->number++;
    }
    public function get()
    {
        return self::$numberStatic . " - " . $this->number;
    }
}
```

We have seen just some very basic examples of the impact on the code if you are going to install and use Laravel Octane. The examples shown earlier were purposely very simple, but with the goal of being easy to understand. In the chapter where we will use Octane in a real scenario, we will cover more realistic examples.

Now we will analyze the impact on performance. So, by installing Octane, what kind of improvement could we have in terms of performance?

Understanding performance measurement in Laravel Octane

We have said that introducing Laravel Octane in your application allows for a performance boost, mainly because the objects and the various instances of the classes used by the framework are no longer initialized at every single HTTP request but at the start of the application server. As a result, for each HTTP request, framework objects are reused. Reusing framework objects saves time in serving the HTTP request.

While, on a logical and understandable level, this can have a positive impact in terms of performance, the goal of this part is to get pragmatic feedback on this performance boost by trying to recover some metrics and values.

In order to provide a rough indication of the benefits and improved response time performance for a request, let us try to perform a simple performance test.

To do this, we are going to install a tool to generate and execute some HTTP-concurrent requests. There are several such tools, one of which is **wrk** (`https://github.com/wg/wrk`).

If you have a macOS environment, you could use the `brew` command (provided by Homebrew) to install the `wrk` tool. To install the tool, use `brew install` as shown:

```
brew install wrk
```

With `wrk`, you can generate concurrent requests for a defined amount of time.

We will conduct two tests for comparison: one test will be conducted with a classical web application on nginx (`http://octane.test`), and the other one with an application served by an application server on Laravel Octane (`http://octane.test:8000`).

The two URLs are resolved as shown:

- `http://octane.test/` is resolved with local address `127.0.0.1` and will reply nginx
- `http://octane.test:8000/` is resolved with local address `127.0.0.1` and port `8000` is bound by Swoole

The `wrk` execution will use 4 threads, open 20 connections, and take 10 seconds of tests.

So, to test NGINX, launch the `wrk` command with these arguments:

```
wrk -t4 -c20 -d10s http://octane.test
```

You will see the following output:

```
Running 10s test @ http://octane.test
  4 threads and 20 connections
  Thread Stats    Avg        Stdev      Max     +/- Stdev
    Latency     51.78ms     61.33ms  473.05ms    88.54%
    Req/Sec    141.79       68.87    313.00      66.50%
  5612 requests in 10.05s, 8.47MB read
  Non-2xx or 3xx responses: 2
Requests/sec:     558.17
Transfer/sec:     863.14KB
```

To test Laravel Octane (RoadRunner), use the following command:

```
wrk -t4 -c20 -d10s http://octane.test:8000
```

You will see the following output:

```
Running 10s test @ http://octane.test:8000
  4 threads and 20 connections
  Thread Stats    Avg        Stdev      Max     +/- Stdev
    Latency    134.58ms    178.24ms   1.09s     79.75%
    Req/Sec    222.72      192.63     1.02k     73.72%
  7196 requests in 10.02s, 8.06MB read
Requests/sec:     718.51
Transfer/sec:     823.76KB
```

This test is very basic because there are no special server-side logic or query databases involved, but it is good to run the test to understand the raw difference in bootstrapping basic objects for Laravel (application container, requests, etc.) and perceive their flavor.

The difference is not so great (7,196 requests versus 5,612 requests) – around 22% – but consider that this difference grows if you add new packages and libraries (more code to be bootstrapped for each request).

Consider also that RoadRunner and Swoole provide other additional tools for improving performances such as enabling concurrency and executing concurrent tasks. The additional tools will be shown later in *Chapters 2* and *3*.

To better explain why Laravel Octane allows you to achieve this improvement, let me demonstrate how and when service providers are instanced and loaded into a service container.

Typically, in a classic Laravel application service, providers are loaded in each request.

Create a new service provider named MyServiceProvider in the app/Providers directory:

```php
<?php

namespace App\Providers;

use Illuminate\Support\ServiceProvider;

class MyServiceProvider extends ServiceProvider
{
    public function __construct($app)
    {
        echo "NEW        - " . __METHOD__ . PHP_EOL;
        parent::__construct($app);
    }
    public function register()
    {
        echo "REGISTER     - " . __METHOD__ . PHP_EOL;
    }
    public function boot()
    {
        echo "BOOT      - " . __METHOD__ . PHP_EOL;
    }
}
```

The new service provider simply shows a message when the service provider is created, registered, and booted.

The lifecycle of a service provider starts with three phases: creation, registration, and boot.

The register() and boot() methods are needed for *dependency resolution*. First of all, every service provider is registered. Once they are all registered, they could be booted. If a service provider needs another service in the boot method, you can be sure that it is ready to be used because it is already registered.

Then, you have to register the service provider, so in `config/app.php` in the `providers` array, add `App\Providers\MyServiceProvider::class`.

In a classical Laravel web application, for every HTTP request, the `MyServiceProvider` service provider is instanced, and the `construct`, `register`, and `boot` methods are called every time, showing this output:

```
NEW           - App\Providers\MyServiceProvider::__construct
REGISTER      - App\Providers\MyServiceProvider::register
BOOT          - App\Providers\MyServiceProvider::boot
```

With Laravel Octane, something different happens.

For a better understanding, we are going to launch the Laravel Octane server with two parameters:

- `workers`: The number of workers that should be available to handle requests. We are going to set this number to 2.

- `max-requests`: The number of requests to process before reloading the server. We are going to set this number to a maximum limit of 5 for each worker.

To start the Octane server with two workers and reload the server after processing five requests, we enter the following command:

```
php artisan octane:start --workers=2 --max-requests=5
```

After launching Octane, try to perform more than one request with the browser accessing this URL: `http://127.0.0.1:8000`.

The following is the output:

```
NEW           - App\Providers\MyServiceProvider::__construct
REGISTER      - App\Providers\MyServiceProvider::register

BOOT          - App\Providers\MyServiceProvider::boot


   200    GET / ...........................................
...... 113.62 ms
NEW           - App\Providers\MyServiceProvider::__construct
REGISTER      - App\Providers\MyServiceProvider::register

BOOT          - App\Providers\MyServiceProvider::boot
```

```
  200       GET / ...............................................................
....... 85.49 ms
  200       GET / ...............................................................
........ 7.57 ms
  200       GET / ...............................................................
........ 6.96 ms
  200       GET / ...............................................................
........ 6.40 ms
  200       GET / ...............................................................
........ 7.27 ms
  200       GET / ...............................................................
........ 3.97 ms
  200       GET / ...............................................................
........ 5.17 ms
  200       GET / ...............................................................
........ 8.41 ms
worker stopped
  200       GET / ...............................................................
........ 4.84 ms
worker stopped
```

The first 2 requests take around 100 **milliseconds** (**ms**), the next requests take 10 ms, and the `register()` and `boot()` methods are called on the first two requests.

So we can see the first two requests (two because we have two workers) are a bit slower (113.62 ms and 85.49 ms) than the next requests (from the third to the tenth request, where we have a response time of less than 10 ms).

Another important thing to mention is that the `register` and `boot` methods are called for the first two requests until the tenth request (two workers multiplied by five max requests). This behavior is repeated for subsequent requests.

And so, installing Laravel Octane in your web application allows you to improve the response time of your application.

All this without having involved certain tools such as concurrency management provided by application servers such as Swoole and RoadRunner.

Summary

Now that we have an overview of the behavior, some benefits, and some side effects of Laravel Octane, we can proceed with the next chapter by installing and configuring one of the two Laravel Octane-compatible application servers: the RoadRunner application server.

We will revisit some of the instructions addressed in this chapter. The goal of this chapter was to provide some useful summary elements to address the more specific and detailed cases in the rest of the book.

Part 2: The Application Server

Laravel Octane uses an application server such as RoadRunner or Swoole. Running the application in an application server instead of a classical web server allows the developer to use some advanced features such as managing multiple workers, concurrent tasks, and bootstrap applications in a more efficient way; fast caching, and sharing data across the workers. This part comprises the following chapters:

- *Chapter 2, Configuring the RoadRunner Application Server*
- *Chapter 3, Configuring the Swoole Application Server*

2
Configuring the RoadRunner Application Server

When developing a web application in Laravel, we are used to using a web server for the delivery of our web application over the network.

A web server exposes the application via the **HTTP** or **HTTPS** protocol and implements functionality that is typically closely related to the delivery of resources via the HTTP protocol.

An application server is a somewhat more structured and complex piece of software in that it can handle different protocols; it can handle HTTP, as well as lower-level protocols such as **TCP**, or other protocols, such as **WebSocket**.

In addition, an application server can implement an articulated worker structure. This means that the application server for the execution of the application logic delegates its execution to a worker. A worker is an isolated thread that is tasked with executing a given task.

Worker management allows applications running via the application server to have access to features such as concurrency and parallel task execution.

The application server to be able to manage the various workers must also be able to implement load distribution features across workers and must also be able to implement proper balancing (using a load balancer).

There are many application servers in the PHP ecosystem, two of which are **RoadRunner** and **Swoole**. They are relevant to the Laravel ecosystem because they are directly supported by Laravel Octane.

These two solutions have different features; however, both allow Laravel Octane to start different workers that will take over the resolution of HTTP requests.

Additional features accessible through Laravel Octane and available only in Swoole (and not in RoadRunner) are the ability to execute multiple concurrent functions, manage shared data in an optimized manner through special tables, and start functions in a scheduled and repetitive mode. We will walk through the additional features provided by Swoole in *Chapter 3, Configuring the Swoole Application Server*.

RoadRunner is perhaps the simplest application server in terms of available features, and it is also the easiest to install.

Therefore, in order to become familiar with the Laravel Octane configuration, we will start with using RoadRunner.

The goal of this chapter is to show you how to set up a basic Laravel application, add Laravel Octane, start using Octane with RoadRunner, and configure it.

Understanding the setup and the configurations are the first steps that allow you to control the behavior of your application.

In this chapter we will cover the following topics:

- Setting up a basic Laravel application
- Installing RoadRunner
- Installing Laravel Octane
- Launching Laravel Octane
- Laravel Octane and RoadRunner advanced configuration

Technical requirements

This chapter will cover the framework and the application server setup (installation and configuration).

The assumption is that you have already installed PHP and Composer. We recommend you have PHP (at least version 8.0) and Composer updated to the latest version.

Typically, we have two main approaches for installing languages and development tools. The first one is to install the tools directly in the operating system of your machine. The second one is to install the tools in an isolated environment such as a virtual machine or Docker.

If you want to follow the instructions and the examples in the book with Docker, the assumption is that you have Docker Desktop already installed on your machine.

For Docker, we will provide you instructions in order to have an image up and running with PHP and Composer.

With this approach, you will be able to run commands and follow examples the same way, regardless of whether you have Docker or want to run PHP and Composer natively.

We will launch commands from the console application (or Terminal Emulator) so the expectation is that you are quite familiar with this type of application (Terminal, iTerm2, Warp for MacOS, Windows Terminal for Windows, Terminator, xterm, GNOME Terminal, Konsole for GNU/Linux, or Alacritty, which is available for all operating systems).

Within the terminal emulator, you will need a shell environment, typically Bash or ZSH (Z shell). We will use the shell configuration to set some environment variables, such as the PATH variable. The PATH variable specifies the directories to be searched to find a command.

> **Source code**
>
> You can find the source code of the examples used in this chapter in the official GitHub repository of this book: `https://github.com/PacktPublishing/High-Performance-with-Laravel-Octane/tree/main/octane-ch02`.

Setting up a basic Laravel application

Our aim in this chapter is to configure Laravel Octane with the RoadRunner application server. To do that, we have to install the RoadRunner application server. However, before that, we must first create a new Laravel application and then add and install the Laravel Octane package. In short, to demonstrate how to install RoadRunner, we will do the following:

1. Create a new Laravel application from scratch.
2. Add the Laravel Octane package to the new Laravel application.
3. Install Laravel Octane, executing a specific command provided by the Laravel Octane package. The command execution will create a basic configuration, which is useful when we start with Laravel Octane. We will show how to install Laravel Octane in the *Installing Laravel Octane* section later.

Getting the Laravel installer

To install Laravel from scratch, you could use the Laravel installer. To globally install the Laravel installer, in your terminal emulator, enter the following:

```
composer global require laravel/installer
```

Once Laravel is installed, be sure that your PATH variable includes the directory where the global composer packages are stored, typically `.composer/vendor/bin/` in your home directory.

To make the PATH variable persistent and to ensure that it is loaded correctly after rebooting your operating system, you can add it to your shell configuration file. For example, if you are using Zshell, add this line in your `.zshrc` file:

```
export PATH=$PATH:~/.composer/vendor/bin/
```

To make sure that your shell configuration is reloaded correctly and you are using Zshell, enter the following:

```
source ~/.zshrc
```

If you have some doubts, restart the console application (the application you are using to launch commands).

To check if everything is fine, try to execute the Laravel installer tool with the `-V` option via command line:

```
laravel -V
```

If you receive an output such as `Laravel Installer 4.2.11`, everything is fine; otherwise, you could see an error such as `command not found`. In this case, my suggestion is to check the following:

- The `laravel` command is present in `~/.composer/vendor/bin/`
- The `laravel` command is executable
- The PATH variable includes the `~/.composer/vendor/bin/` directory

To check whether the Laravel installer command is present and executable, you can check it with the classic `ls` command:

```
ls -l ~/.composer/vendor/bin/laravel
```

And to see if the permissions include the `x` char, you will see something like `-rwxr-xr-x`.

If the command exists in the right place without executable permission, you can fix it with the `chmod` command, adding executable (`+x`) permission to the owner (`u`):

```
chmod u+x ~/.composer/vendor/bin/laravel
```

If the command exists and has the right permissions, check if the PATH variable is correct and includes the `~/.composer/vendor/bin/` path.

If the PATH variable doesn't include the right path, check that you added it to the PATH variable and if the PATH variable includes the right path, be sure to have reloaded the shell environment or, at least, restart your terminal emulator.

I wanted to spend a few additional words on this type of check. This type of check is useful and will continue to be useful as we add new commands. The existence of the command, its permissions, and its reachability are checks that can save time if we run into problems running a newly installed command.

Now, let me show you how to install a Laravel application before adding Laravel Octane.

Installing a new Laravel web application from scratch

To create a new basic Laravel application, we can use the Laravel installer:

```
laravel new octane-ch2
```

If you don't have a Laravel installer, you can use the composer command to install the Laravel application:

```
composer create-project laravel/laravel octane-ch2
```

In this basic usage, these commands (laravel new and composer create-project) are pretty similar. They do the following:

- Clone the laravel/laravel repository

- Create a .env file from .env.example

- Install all dependencies found in composer.json

- Generate optimized autoload files

- Register or discover any new supported packages via the execution of the php artisan package:discover command

- Publish laravel-assets files

- Generate the application key in the .env file

I suggest you use the Laravel command because it has some additional options and arguments that allow you to enable some nice features, such as adding Jetstream scaffolding, choosing the Livewire stack or Inertia for Jetstream, and enabling teams management for Jetstream. All these options, for the goal we currently have, are not needed, so for this reason using the first or second command has the same result.

So, now you can enter the new octane-ch2 directory to check your new Laravel application.

To launch the internal web server, you can use the artisan command provided by Laravel:

```
php artisan serve
```

If you open the browser at http://127.0.0.1:8000 you can see the default home page of Laravel.

Now that we have our Laravel application up and running, it's time to set up the RoadRunner application server.

Installing RoadRunner

RoadRunner is a PHP application server that is mature and stable, so you can use it in your production environment. It is written in the **Go** programming language, which means that under the hood, RoadRunner uses goroutines and the multi-threading capabilities provided by Go. Thanks to its Go implementation, RoadRunner runs on the most common operating systems, such as macOS, Windows, Linux, FreeBSD, and ARM.

Thanks again to its Go implementation, RoadRunner is released as a binary, so the installation process is very simple.

RoadRunner is an open source project, so you have access to the source code, binaries, and documentation:

- The source code: `https://github.com/roadrunner-server/roadrunner`
- The binary releases: `https://github.com/roadrunner-server/roadrunner/releases`
- The main documentation: `https://roadrunner.dev/docs/readme/2.x/en`

We can obtain RoadRunner in more than one way.

Just for starting quickly, I'm going to use the Composer method. The Composer approach requires two steps:

1. Install the RoadRunner CLI.
2. Get RoadRunner binaries via the RoadRunner CLI.

So, as a first step, let me install the RoadRunner CLI, according to the official documentation available at `https://roadrunner.dev/docs/intro-install`:

```
composer require spiral/roadrunner:v2.0 nyholm/psr7
```

As you can see, we are going to add two packages:

- RoadRunner CLI version 2
- The Nyholm implementation of **PSR7**

Nyholm

Nyholm is an open source PHP package that implements the PSR7 standard. The source code is here: `https://github.com/Nyholm/psr7`.

At the end, the `composer` command adds two lines into the `require` section of your `composer.json` file:

```
"nyholm/psr7": "^1.5",
"spiral/roadrunner": "v2.0"
```

The mentioned, PSR7 is a standard that defines the PHP interfaces for representing HTTP messages and URIs. In this way, if you are going to use a library to manage HTTP messages and URIs, and the library implements a PSR7 standard, you know that you have some methods with a standardized signature to manage the request. For example, you know that you have `getMethod()` to retrieve the HTTP method, and the value is a string (because is defined by the standard).

After installing the RoadRunner CLI via Composer, you will find the `rr` executable in the `vendor/bin/` directory.

To check that it's there, use this command:

```
ls -l vendor/bin/rr
```

You will see a file, more or less 3 KB, with executable permissions (denoted by the `x` symbol).

This executable is the RoadRunner CLI, which allows you to install the RoadRunner application server executable. To obtain the executable, you can execute the RoadRunner CLI with the `get-binary` option:

```
vendor/bin/rr get-binary
```

The output generated by the command execution will show the version of the package, the operating system, and the architecture as shown in the figure below:

```
→  octane-ch2 git:(main) ✗ vendor/bin/rr get-binary

Environment:
   - Version:            2.*
   - Stability:          stable
   - Operating System:   darwin
   - Architecture:       arm64

   - roadrunner-server/roadrunner (v2.10.5): Downloading...
RoadRunner (v2.10.5) has been installed into /Users/roberto/Sites/octane-ch2/rr
```

Figure 2.1: Getting the RoadRunner executable

You might have some confusion in that the RoadRunner CLI executable is called `rr`, like the RoadRunner executable. What you should know is that the RoadRunner CLI is stored in the `vendor/bin` directory, while the RoadRunner application server executable is stored in the project root directory. Also, the CLI is about 3 KB, while the application server is about 50 MB.

```
→  octane-ch2 git:(main) × ls -lh rr vendor/bin/rr
-rwxr-xr-x  1 roberto   staff     49M Jul  3 09:45 rr
-rwxr-xr-x  1 roberto   staff    3.1K Jul  3 09:13 vendor/bin/rr
```

Figure 2.2: The two rr executables, the CLI and the application server

In addition, you can run the two executables with the option to show the version:

```
→  octane-ch2 git:(main) × ./rr -v
rr version 2.10.5 (build time: 2022-06-23T21:11:20+0000, go1.18.3)
→  octane-ch2 git:(main) × ./vendor/bin/rr -V
RoadRunner CLI 2.0.0
```

Figure 2.3: The rr versions

Now that we have installed the `rr` executables (RoadRunner), we can start to use it.

Executing the RoadRunner application server (without Octane)

To execute the RoadRunner application server with a basic example, we need to do the following:

- Create a configuration file
- Create a PHP script called by the application server once an HTTP request hits the application server
- Launch the application server

The configuration file for RoadRunner by default is `.rr.yaml`. It has a lot of configuration directives and parameters.

A minimal configuration file requires a few things:

- The command to launch for each worker instance (`server.command`)
- The address and port to bind and listen for new HTTP connections (`http.address`)
- The number of workers to launch (`http.pool.num_workers`)
- The level of the log (`logs.level`)

An example of a configuration file with the preceding considerations is shown here:

```
version: '2.7'

server:
  command: "php test-rr.php"

http:
  address: "0.0.0.0:8080"
  pool:
    num_workers: 2

logs:
  level: info
```

With this configuration file, `test-rr.php` is the script to launch for workers, `8080` is the port to listen to connections, with `2` workers and `info` for the log levels.

The script file for implementing the logic of the workers is `test-rr.php`:

```php
<?php

include 'vendor/autoload.php';

use Nyholm\Psr7;
use Spiral\RoadRunner;

$worker = RoadRunner\Worker::create();
$psrFactory = new Psr7\Factory\Psr17Factory();
$worker = new RoadRunner\Http\PSR7Worker($worker, $psrFactory,
$psrFactory, $psrFactory);
// creating a unique identifier specific for the worker
$id = uniqid('', true);
echo "STARTING ${id}";

while ($req = $worker->waitRequest()) {
    try {
        $rsp = new Psr7\Response();
```

```
        $rsp->getBody()->write("Hello ${id}");
        echo "RESPONSE SENT from ${id}";
        $worker->respond($rsp);
    } catch (\Throwable $e) {
        $worker->getWorker()->error((string) $e);
        echo 'ERROR ' . $e->getMessage();
    }
}
```

The script does the following:

- Includes vendor/autoload.php
- Instances the worker object with classes provided by RoadRunner (RoadRunner\Http\PSR7Worker)
- Generates a unique ID for showing how the traffic is balanced and delegated to the two workers ($id = uniqid('', true))
- Waits for a new connection ($worker->waitRequest())
- Generates a new response ($worker->respond()) when a new connection request arrives

With the configuration file and the preceding worker script, you can launch the application server with the serve option:

./rr serve

With this configuration, you will see one server started and two workers started by the server.

Now you can hit the serve via the curl command. The curl command is a command that sends an HTTP request to a specific URL.

In another instance for the terminal emulator (or another tab), launch the following:

curl localhost:8080

By executing curl four times, we will send four different requests to the application server to port 8080.

On the terminal emulator, if you launch the application server, you will see the log of the application server:

```
→  octane-ch2 git:(main) ✗ ./rr serve
[INFO] RoadRunner server started; version: 2.10.5, buildtime: 2022-06-23T21:11:20+0000
2022-07-03T21:57:14.548+0200    INFO    server          STARTING 62c1f49a85d525.63335299
2022-07-03T21:57:14.548+0200    INFO    server          STARTING 62c1f49a85d505.45719503
2022-07-04T18:14:09.925+0200    INFO    server          RESPONSE SENT from 62c1f49a85d505.
45719503
2022-07-04T18:14:09.928+0200    INFO    http            http log        {"status": 200, "m
ethod": "GET", "URI": "/", "remote_address": "127.0.0.1:63560", "read_bytes": 0, "write_by
tes": 29, "start": "2022-07-04T18:14:09.909+0200", "elapsed": "19.214042ms"}
2022-07-04T18:14:25.467+0200    INFO    server          RESPONSE SENT from 62c1f49a85d525.
63335299
2022-07-04T18:14:25.467+0200    INFO    http            http log        {"status": 200, "m
ethod": "GET", "URI": "/", "remote_address": "127.0.0.1:63625", "read_bytes": 0, "write_by
tes": 29, "start": "2022-07-04T18:14:25.462+0200", "elapsed": "5.028542ms"}
2022-07-04T18:14:31.346+0200    INFO    server          RESPONSE SENT from 62c1f49a85d505.
45719503
2022-07-04T18:14:31.346+0200    INFO    http            http log        {"status": 200, "m
ethod": "GET", "URI": "/", "remote_address": "127.0.0.1:63650", "read_bytes": 0, "write_by
tes": 29, "start": "2022-07-04T18:14:31.346+0200", "elapsed": "349.25µs"}
2022-07-04T18:33:10.487+0200    INFO    server          RESPONSE SENT from 62c1f49a85d525.
63335299
2022-07-04T18:33:10.488+0200    INFO    http            http log        {"status": 200, "m
ethod": "GET", "URI": "/", "remote_address": "127.0.0.1:49902", "read_bytes": 0, "write_by
tes": 29, "start": "2022-07-04T18:33:10.487+0200", "elapsed": "805.5µs"}
```

Figure 2.4: The INFO log of the application server

The most relevant thing is that after the first two requests, the elapsed time is reduced by at least an order of magnitude.

If you look at the `elapsed` value, you will see that the first request takes 20 milliseconds to be executed, while subsequent requests take approximately some hundred microseconds (1 millisecond is equivalent to 1,000 microseconds).

The response time in milliseconds (the absolute value) probably depends on multiple factors (load, resources, memory, CPU). Take a look at the relative value and how much the response time decreases in the next requests. The response times are dramatically reduced from a few milliseconds to a few microseconds.

So, we are saying that thanks to the architecture based on workers implemented with RoadRunner, we can improve the performance, especially for requests after the first request.

But how do we include and use RoadRunner in our Laravel application?

The preceding example uses objects and methods shipped by RoadRunner in a pure PHP environment. Now we have to figure out how all of these features/improvements could be included in Laravel, especially for all things related to the bootstrap of the framework.

This is the goal of Octane. It allows us to use the RoadRunner features while hiding the complexity of the integration, the bootstrap process, and the configuration.

Installing Laravel Octane

The script file (`test-rr.php`) and the configuration file (`.rr.yaml`) were created so that the dynamics of the operation of RoadRunner can be understood. Now, let's focus on the installation of Laravel Octane. Let's pick up the discussion from the installation of the Laravel application via the `laravel new` command and the installation of the RoadRunner executable by running `composer require` and then running the `rr get-binaries`. Let me recap quickly:

```
# installing Laravel application
laravel new octane-ch2b
# entering into the directory
cd octane-ch2b
# installing RoadRunner CLI
composer require spiral/roadrunner:v2.0 nyholm/psr7
# Obtaining Roadrunner Application Server executable, via CLI
vendor/bin/rr get-binary
```

Now you can install Laravel Octane:

```
composer require laravel/octane
```

Then you can correctly configure Laravel Octane with the `octane:install` command:

```
php artisan octane:install
```

With the latest command, you have to decide whether to use RoadRunner or Swoole. For the purpose of this chapter, select RoadRunner. We will cover Swoole in the next chapter.

The tasks performed by `octane:install` are as follows:

- Avoiding commit/push on Git repository RoadRunner files: check and eventually fix the `gitignore` file that includes `rr` (the RoadRunner executables) and `.rr.yaml` (the RoadRunner configuration file).

- Ensuring the RoadRunner package is installed in the project. If not, it executes the `composer require` command.

- Ensuring the RoadRunner binary is installed into the project. If not, it executes `./vendor/bin/rr get-binary` to download the RoadRunner application server.

- Ensuring the RoadRunner binary is executable (`chmod 755`).

- Checking some requirements, such as version 2.x, if the RoadRunner application server has already been installed.

- Setting the OCTANE_SERVER environment variable in the .env file (if it's not already present).

The last octane:install command will create a config/octane.php file and will also add a new configuration key to the .env file. The new key is named OCTANE_SERVER and the value is set to roadrunner.

This value is used in the config/octane.php file:

```
return [

    /*
    |--------------------------------------------------------------------
    | Octane Server
    |--------------------------------------------------------------------
    |
    | This value determines the default "server" that will
      be used by Octane
    | when starting, restarting, or stopping your server
      via the CLI. You
    | are free to change this to the supported server of
      your choosing.
    |
    | Supported: "roadrunner", "swoole"
    |
    */

    'server' => env('OCTANE_SERVER', 'roadrunner'),
```

So, with the environment variable, you can control which application server you want to use.

Now that we have installed Laravel Octane, it's time to launch it.

Launching Laravel Octane

To launch Laravel Octane, run the following command:

```
php artisan octane:start
```

Once Laravel Octane is started you can visit http://127.0.0.1:8000 in your browser.

Your browser will be shown the classic Laravel welcome page. There are no visual differences in the welcome pages of Laravel and Laravel Octane. The big difference is the way your application is served via HTTP.

You can control the execution of Octane with some parameters:

- --host: Default 127.0.0.1, the IP address the server should bind to
- --port: Default 8000, the port the server should be available on
- --workers: Default auto, the number of workers that should be available to handle requests
- --max-requests: Default 500, the number of requests to process before reloading the server

For example, you can launch Octane with just two workers:

```
php artisan octane:start --workers=2
```

So now, open the page at http://localhost:8000 more than twice (two is the number of workers). You can open the page via your browser or by launching curl:

```
curl localhost:8000
```

You can see something that we already know because of the previous tests with RoadRunner installed without Laravel. The first two requests (two is the number of workers) are slower than the next requests.

The following output is related to the log shown by the Laravel Octane server:

```
200     GET /  ..................................... 76.60 ms
200     GET /  ..................................... 60.39 ms
200     GET /  ..................................... 3.46 ms
200     GET /  ..................................... 2.70 ms
200     GET /  ..................................... 2.66 ms
200     GET /  ..................................... 3.66 ms
```

If you launch the server, define the maximum number of requests to process (for each worker) before starting the server with the max-requests parameter:

```
php artisan octane:start --workers=2 --max-requests=3
```

You can see a similar output but after six requests (a maximum of three requests for two workers), you will see that the message worker has stopped and the response after the stopped worker takes the same amount of time as the first two requests:

```
200      GET / ................................. 86.56 ms
200      GET / ................................. 52.30 ms
200      GET / .................................. 2.38 ms
200      GET / .................................. 2.73 ms
200      GET / .................................. 2.57 ms
worker stopped
200      GET / .................................. 2.75 ms
worker stopped
200      GET / ................................. 63.95 ms
200      GET / ................................. 60.83 ms
200      GET / .................................. 1.75 ms
200      GET / .................................. 2.74 ms
```

Why is restarting the server important? To ensure we prevent any memory leak issues due to the long life cycle of the objects (the server and the workers), it is a common practice to reset the status. If you are not going to define the max-requests parameter in the command line, the default is set automatically by Laravel Octane to 500.

In the classic web server scenario (without Laravel Octane), the life cycles of all objects related to your application, but especially to the objects automatically instantiated and managed by the framework, are confined to each individual request. In every single request, all the objects necessary for the framework to work are instantiated, and the objects are destroyed when the response is sent back to the client. This also explains and helps you understand why the response times in a classic framework with a web server are longer than the response times of an already initialized worker.

Now that Laravel Octane is launched, we can look at its configuration.

Laravel Octane and RoadRunner advanced configuration

As mentioned in the previous section, we can control some parameters during the launch of Laravel Octane. This is because you want to change some options, such as the number of workers or the port and, like in the next examples, if you want to activate the HTTPS protocol.

Under the hood, Octane collects parameters from the command line and some Octane configuration and starts the RoadRunner process (it starts the rr command).

In the Octane source code, there is a file called `StartRoadRunnerCommand.php` that implements a Laravel artisan command with the following code:

```php
$server = tap(new Process(array_filter([
    $roadRunnerBinary,
    '-c', $this->configPath(),
    '-o', 'version=2.7',
    '-o', 'http.address='.$this->option('host').':
      '.$this->option('port'),
    '-o', 'server.command='.(new PhpExecutableFinder)-
      >find().' '.base_path(config('octane.roadrunner
      .command', 'vendor/bin/roadrunner-worker')),
    '-o', 'http.pool.num_workers='.$this->workerCount(),
    '-o', 'http.pool.max_jobs='.$this->option(
      'max-requests'),
    '-o', 'rpc.listen=tcp://'.$this->option('host').':
      '.$this->rpcPort(),
    '-o', 'http.pool.supervisor.exec_ttl='
      .$this->maxExecutionTime(),
    '-o', 'http.static.dir='.base_path('public'),
    '-o', 'http.middleware='.config(
      'octane.roadrunner.http_middleware', 'static'),
    '-o', 'logs.mode=production',
    '-o', app()->environment('local') ? 'logs.level=debug'
      : 'logs.level=warn',
    '-o', 'logs.output=stdout',
    '-o', 'logs.encoding=json',
    'serve',
]), base_path(), [
    'APP_ENV' => app()->environment(),
    'APP_BASE_PATH' => base_path(),
    'LARAVEL_OCTANE' => 1,
]))->start();
```

Looking at this source code helps you understand which parameters are used to launch the RoadRunner executable.

With the `-c` option (`$this->configPath()`), an additional configuration file is loaded. This means that if there are no basic options managed by Octane that match your expectations, you can define them in the `.rr.yaml` configuration file.

The basic parameters managed by Octane (shown in the previous section) are hostname, port, worker count, max requests, max execution time for supervisor, the HTTP middleware, and the log level.

The RoadRunner configuration file allows you to load special and advanced configurations. A classic example is the option to allow local RoadRunner instances to listen to and receive HTTPS requests.

Why do you need to serve HTTPS locally in the development environment? You might need to activate the HTTPS protocol because some browser features are available only when the page is served via HTTPS or localhost. These features are geolocation, device motion, device orientation, audio recording, notifications, and so on.

Typically, during local development, we are used to serving pages via localhost. In this context, there is no need to serve the traffic via HTTPS. However, if we wanted to expose pages to the local network so that we could test our web application via a mobile device connected to the local network, we would have to make sure that the service is reachable via a valid local network address and thus localhost is not sufficient. In this case (for those special browser features), HTTPS is needed.

Or another scenario, wherein your page served locally is included in a web page (via iFrame or as assets) and the host page is served via HTTPS. In this case, including assets or including a page delivered by HTTP in an HTTPS context raises a security exception in the browser.

If you want to configure Octane to serve HTTPS requests, you have to do the following:

- Install a tool such as **mkcert** that allows you to create and manage certificates. Because of the design and implementation of HTTPS, the protocol requires public/private certificates to work.
- Create certificates for localhost or addresses you want.
- See the location of the CA certificate and key storage location.

To better understand what is needed, let's look at RoadRunner's configuration for HTTPS:

```
version: "2.7"

http:
  # host and port separated by semicolon
  address: 127.0.0.1:8000

  ssl:
```

```
# host and port separated by semicolon (default :443)
address: :8893
redirect: false
# Path to the cert file. This option is required for
# SSL working.
# This option is required.
cert: "./localhost.pem"
# Path to the cert key file.
# This option is required.
key: "./localhost-key.pem"
# Path to the root certificate authority file.
# This option is optional.
root_ca: "/Users/roberto/Library/Application\
          Support/mkcert/rootCA.pem"
```

The two mandatory fiels and one optional file are as follows:

- Cert: The cert file
- Key: The cert key file
- Root_ca: The root certificate authority file

With the first two files, HTTPS works but a warning is raised by your browser (there's not a valid certificate, because the certificate is self-signed). Filling only the two first parameters, the certificate is evaluated as self-signed and, typically, browsers tend not to consider such certificates trustable.

With the third file, the browser allows you to browse via HTTPS without any warnings (the certificate is valid).

So, first, you have to install **mkcert**. The Git repository of mkcert is https://github.com/FiloSottile/mkcert.

mkcert is an open source tool available for all platforms.

The instructions to install mkcert and create certificates for macOS are as follows:

```
brew install mkcert
mkcert -install
mkcert localhost
```

If you are using Windows, you could use the Chocolatey package manager (`https://chocolatey.org/`) and use the following command:

```
choco install mkcert
```

For GNU/Linux, you can use the package manager provided by your distribution.

Now you have two new files in the project directory: `localhost-key.pem` and `localhost.pem`.

> **Note**
>
> I strongly recommend listing these two files in the `.gitignore` file prevent them from being pushed into your Git repository (if it is used).

You can use the first one in your `.rr.yaml` file for the `key` parameter and the second one for the `cert` parameter.

To fill the `root_ca` parameter, you have to see where CA files are stored via the `mkcert` command (with the `CAROOT` option):

```
mkcert -CAROOT
```

This command will show you the directory where the CA files are stored.

To see the name of CA files, run the following:

```
ls "$(mkcert -CAROOT)"
```

You can fill the `root_ca` parameter with the full path of the `rootCA.pem` file.

> **Note**
>
> If you are using Firefox and you are still getting the self-signed certificate warning, install `certutil` (with Homebrew, `certutil` is included in the `nss` package, so execute `brew install nss`) and then execute `mkcert -install` again (and restart the Firefox browser).

Now you can start Octane with the following:

```
php artisan octane:start
```

Open your browser at the URL defined in the `address` parameter According to the parameters used in the last example (the RoadRunner configuration in the `.rr.yaml` file), you should open your browser and open the page at this URL: `https://127.0.0.1:8893`. (note that `https://` is the protocol instead of `http://`)

So, now you are familiar with installing RoadRunner with Laravel Octane, starting the Octane server, and accessing the advanced configuration.

Summary

In this chapter, we explored the installation and configuration of Laravel Octane with the RoadRunner application server. We took a look at the benefit we get from the usage of RoadRunner and how to enable advanced features such as the HTTPS protocol.

In the next chapter, we will see how to do the same things with Swoole. We will see the additional features Swoole has, compared with RoadRunner, and in *Chapter 4*, *Building a Laravel Octane Application*, we will start to look at the code for the web application using the Octane service, which is now up and running.

3

Configuring the Swoole Application Server

With Laravel Octane, instead of RoadRunner, you can use another type of application server. In addition to RoadRunner, we can configure and use Swoole. Both tools allow you to implement an application server. Obviously, there are differentiating elements between the two tools, and sometimes deciding which one to use can be difficult.

As we saw, RoadRunner is a separate executable, which means that its installation, as seen in *Chapter 2, Configuring the RoadRunner Application Server*, is quite simple and does not affect the heart of the PHP engine. This means that in all probability, it has no side effects on other tools that work in the heart of the engine. These tools (such as Xdebug) are typically diagnostic, monitoring, and debugging tools.

The advice I would like to give is that, in the process of analysis and selection of the application server, evaluate the various other tools that may be useful in the development process (such as Xdebug) and evaluate their compatibility with Swoole.

Despite there being a greater complexity in the management of Swoole, Swoole offers benefits such as additional and advanced features, such as the management of an in-memory cache (benefits on performance), Swoole Table, which is a data store for sharing information between different processes (and facilitates information sharing and better cooperation between processes), and the ability to start asynchronous functions and processes at a specific interval (thus enabling asynchronous and parallel execution of functions).

In the chapter, we will see how to install and configure Swoole with Laravel Octane and how to best use the specific features provided by Swoole. In this chapter, we will explore features of Swoole such as executing concurrent tasks, executing interval tasks, using the highly performant cache and storage mechanisms provided by Swoole, and accessing metrics about the usage of workers. All these features are available via Octane thanks to the Swoole application server.

In particular, we will cover the following:

- Setting up Laravel Octane with Swoole using Laravel Sail

- Installing Open Swoole

- Exploring Swoole features

Technical requirements

This chapter will cover the Swoole and Open Swoole application server setup (installation and configuration).

Unlike what we did for RoadRunner, in this case, we have to install a **PHP Extension Community Library** (**PECL**) extension to allow PHP to be able to operate with Swoole.

> **What is PECL?**
> PECL is the repository for PHP extensions. The PECL extensions are written in C language and have to be compiled to be used with the PHP engine.

Because of the complexity of installing the PECL module with all the dependencies required by the compilation, configuration, and setup of the PHP extension, we are going to use a container approach – so instead of installing all the necessary tools for compiling the PHP extension on our personal operating system, we will use **Docker**.

This allows us to have a running operating system (container) hosted within our real operating system. The purpose of having an isolated operating system in a container is to contain everything you need for the PHP development environment.

This means that if we install dependencies and tools, we will do it within an isolated environment without affecting our real operating system.

In the previous chapter, we did not use this approach as the requirements were simpler. However, many people use the container method for every development environment, even the simplest ones.

As the development environment begins to require additional dependencies, it may become unmanageable. It could become especially unmanageable in the case of evolution: try to think of the simultaneous management of several versions of PHP, which may require additional versions of dependencies. To isolate and limit all this within a consistent environment, it is recommended to use a container approach. To do this, the installation of **Docker Desktop** (https://www.docker.com/products/docker-desktop/) is suggested.

Once we have installed Docker Desktop, we will configure a specific image for PHP with the extensions we need.

All these new packages that we are going to install will be stored within this image. When we delete the development environment, simply delete the image used. The only tool installed on our operating system will be Docker Desktop.

So, to install Docker Desktop, simply download and proceed with the installation wizard specific to your operating system. In the case of macOS, please refer to the type of reference chip (Intel or Apple).

If you do not have a thorough knowledge of Docker, do not worry – we are going to use another powerful tool in the Laravel ecosystem: **Laravel Sail**.

Laravel Sail is a command-line interface that exposes commands for managing Docker, specifically for Laravel. Laravel Sail simplifies the use and configuration of Docker images, allowing the developer to focus on the code.

We are going to use Laravel Sail commands for the creation of a development environment, but under the hood, they will result in Docker commands and ready-to-use Docker configurations.

> **Source code**
>
> You can find the source code of the examples used in this chapter in the official GitHub repository of this book: `https://github.com/PacktPublishing/High-Performance-with-Laravel-Octane/tree/main/octane-ch03`.

Setting up Laravel Octane with Swoole using Laravel Sail

In order to have an environment up and running with Swoole as the application server and using a Docker container, you have to follow some steps:

1. Set up Laravel Sail
2. Install Laravel Octane
3. Set up Laravel Octane and Swoole

Setting up Laravel Sail

First, create your Laravel application:

```
laravel new octane-ch03
cd octane-ch03
```

Or, if you already have your Laravel application you can use the `composer show` command to check whether Laravel Sail is installed. This command also shows you some additional information about the package:

```
composer show laravel/sail
```

If Laravel Sail is not installed, run `composer require laravel/sail --dev`.

Once you have Sail installed, you have to create a `docker-compose.yml` file. To create a `docker-compose.yml` file, you can use the `sail` command, `sail:install`:

```
php artisan sail:install
```

The `sail:install` command will create Docker files for you. The `sail:install` process will ask you which service you want to enable. Just to start, you can select the default item (`mysql`):

```
→ octane-ch3 php artisan sail:install

Which services would you like to install? [mysql]:
  [0] mysql
  [1] pgsql
  [2] mariadb
  [3] redis
  [4] memcached
  [5] meilisearch
  [6] minio
  [7] mailhog
  [8] selenium
> 0

Sail scaffolding installed successfully.
```

Figure 3.1 – The Laravel Sail services

Answer the questions in the `sail:install` command to determine which services to include. The Sail configuration for Docker starts from a set of ready-to-use templates (stubs), and the `sail:install` command includes the necessary ones. If you are curious about which templates are included and how they are implemented, look here: `https://github.com/laravel/sail/tree/1.x/stubs`.

If you take a look at the templates, you will see that they make use of environment variables such as `${APP_PORT:-80}`. This means that you can control the configuration through environment variables, which are configurable through the `.env` file. The `.env` file is automatically generated by the installation of Laravel. If for some reason the `.env` file is not present, you can copy the `.env` file from the `.env.example` file (for example, when you clone an existent repository that uses Laravel Octane, probably the `.env` file is included in the `.gitignore` file). In the example, if you want to customize the port where the web server receives the request (`APP_PORT`), simply add the parameter to the `.env` file:

```
APP_PORT=81
```

In this case, the web server that will serve your Laravel application will do so via port 81. The default is 80 as seen from ${APP_PORT:-80}.

> **Note**
>
> The examples in the *Using Swoole features* section will use Laravel Sail with APP_PORT set to 81, so all examples will refer to the host http://127.0.0.1:81

If you have made this change to the .env file, you can now launch the command from your Laravel project directory:

```
./vendor/bin/sail up
```

This will start the Docker container. The first time, the execution may take some time (a few minutes) because the preconfigured images with nginx, PHP, and MySQL will be downloaded.

Once the execution of the command is complete, you can visit the http://127.0.0.1:81 page and see your Laravel welcome page:

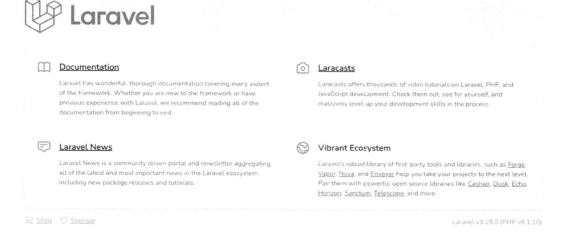

Figure 3.2 – Laravel welcome page

The main difference with this container approach is that the tools used to render the page (nginx, PHP) are included in the Docker image and not in your main operating system. With this method, you may not even have the PHP and nginx engine on your main operating system. The Laravel Sail configuration instructs Docker to use the container for PHP and all the needed tools, and points to your local source code (the root of your Laravel project) on your local filesystem.

Now that we have Laravel Sail installed with your Laravel application, we have to add Laravel Octane stuff to use the Swoole package already included in the Laravel image provided by Sail.

Let's begin by installing Laravel Octane.

Installing Laravel Octane

We are going to install Laravel Octane through the container provided by Laravel Sail.

So while `sail up` is still running (is running as a server), launch `composer require` with the Octane package. Launching the command with `sail` will execute your command inside your container:

```
./vendor/bin/sail composer require laravel/octane
```

Setting up Laravel Octane and Swoole

In order to adjust the command that launches the server, you have to publish Laravel Sail files with the following command:

```
./vendor/bin/sail artisan sail:publish
```

This command will copy the Docker configuration files from the package in the `docker` directory in the root project directory.

It creates a `docker` directory. Inside the `docker` directory, more than one directory is created, one for each PHP version: `7.4`, `8.0`, `8.1`.

In the `docker/8.1` directory, you have the following:

- A Dockerfile.
- The `php.ini` file.
- The `start-container` script used to launch the container. This script refers to the `supervisor.conf` file with the bootstrap configuration.
- The `supervisor.conf` file with the configuration of the supervisor script.

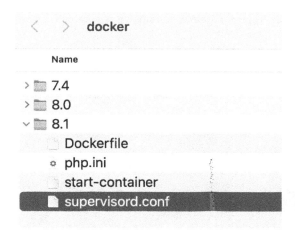

Figure 3.3 – The Docker configuration files (after sail:publish execution)

The `supervisord.conf` file is important because it includes the bootstrap command of the web server of the container. By default, the `supervisord.conf` file contains the command directive:

```
command=/usr/bin/php -d variables_order=EGPCS /var/www/html/
artisan serve --host=0.0.0.0 --port=80
```

Now, we have Laravel Octane, so instead of using the classic `artisan serve`, we have to change it and use `artisan octane:start`; so, in the `docker/supervisor.conf` file, you have to adjust the command line:

```
command=/usr/bin/php -d variables_order=EGPCS /var/www/html/
artisan octane:start --server=swoole --host=0.0.0.0 --port=80
```

If you look, you will see that with `artisan octane:start`, the Swoole server with the `--server=swoole` parameter is also defined.

> **Note**
>
> If you are confused by ports `80` and `81`, just to clarify: the application server inside the container will listen internally to port `80`. With `APP_PORT=81`, we are instructing Docker to map the external connections from port `81` to internal port `80`.

When you change some Docker configuration files, you have to rebuild the image in order for the container to use the changed files. To rebuild the image, use the `build` option:

```
./vendor/bin/sail build --no-cache
```

This command takes a while to be completed, but once it is completed, you can execute the `sail` up command:

```
./vendor/bin/sail up
```

When Octane is ready, you will see the `INFO   Server running...` message:

```
octane-ch03-laravel.test-1  |
octane-ch03-laravel.test-1  |    INFO  Server running...
octane-ch03-laravel.test-1  |
octane-ch03-laravel.test-1  |   Local: http://0.0.0.0:80
octane-ch03-laravel.test-1  |
octane-ch03-laravel.test-1  |   Press Ctrl+C to stop the server
octane-ch03-laravel.test-1  |
```

Figure 3.4 – The Octane server is running

If you open your browser, you can access `http://localhost:81`. The message in your console says that the server is listening to port `80`, but as we previously mentioned, the process wherein the server is listening to port `80` is the internal process inside the Docker container. The processes external to the Docker container (your browser) have to refer to the exposed port (`81` according to the `APP_PORT` configuration).

In your browser, you will see the default welcome Laravel page. You can tell that this page is served by Octane and Swoole from the HTTP server header from the response. A way to see this is to launch the `curl` command with the `-I` option to show the header, and filter the header needed with the `grep` command:

```
curl -I 127.0.0.1:81 -s | grep ^Server
```

The output will be as follows:

```
Server: swoole-http-server
```

This means that the Laravel application is served by Octane and the Swoole application server.

So, we can start to use some Swoole functionalities – but before that, let us install Open Swoole.

Installing Open Swoole

Laravel Sail uses by default an image with PHP that includes Swoole modules. Swoole is distributed as a PECL module and you can find it here: `https://pecl.php.net/package/swoole`. The sources are here: `https://github.com/swoole/swoole-src`.

Some developers forked the source of Swoole, creating the Open Swoole project to address security concerns.

The reason for the fork is reported here: `https://news-web.php.net/php.pecl.dev/17446`.

So, if you want to use Swoole as the engine for Laravel Octane, you could decide to use the Open Swoole implementation. If you want to use Open Swoole, the installation and the configuration are the same as Swoole; Open Swoole is also distributed as a PECL module.

Laravel Octane supports both.

For demonstration purposes, I will install Open Swoole for a new Laravel project directly in the operating system (no Docker):

```
# installing new Laravel application
laravel new octane-ch03-openswoole
# entering into the new directory
cd octane-ch03-openswoole
# installing Pecl module
pecl install openswoole
# installing Octane package
composer require laravel/octane
# installing Laravel Octane files
php artisan octane:install
# launching the OpenSwoole server
php artisan octane:start
```

To check that the HTTP response is created by the Open Swoole server, in another terminal session, launch the following `curl` command:

```
curl -I 127.0.0.1:8000 -s | grep ^Server
```

Here's the output:

```
Server: OpenSwoole 4.11.1
```

So, Open Swoole is a fork of the Swoole project. We will refer to it as Swoole; you can decide which you want to install. The features discussed in the *Exploring Swoole features* section are supported both by Swoole and Open Swoole.

Before exploring the Swoole features, we should install one more package that improves the developer experience.

Before editing the code

We are going to use Swoole functionalities, implementing some example code. When you change (or edit) your code and Laravel Octane has already loaded the worker, you have to reload the worker. Manually, you can use an Octane command. If you are using Laravel Sail (so Docker), you have to run the command in the container. The command is as follows:

```
php artisan octane:reload --server=swoole
```

If you are running in a container, you have to use the `sail` command:

```
vendor/bin/sail php artisan octane:reload --server=swoole
```

If you want to avoid manually reloading workers every time you edit or change your code and you want that Octane watches automatically the file changes, you have to do the following:

- Install the **chokidar** node package used by Octane in `watch` mode
- Change the `supervisord` configuration file in order to launch Octane with the `--watch` option
- Rebuild the image to reflect the changes
- Execute `Sail` again

So, first, let's install `chokidar`:

```
npm install --save-dev chokidar
```

In `docker/8.1/supervisord.conf`, add the `--watch` option to the `octane:start` command directive:

```
command=/usr/bin/php -d variables_order=EGPCS /var/www/html/
artisan octane:start --server=swoole --host=0.0.0.0 --port=80
--watch
```

Then, rebuild the Docker image to be sure that the update to the configuration takes effect and then launch the Laravel Sail service:

```
vendor/bin/sail build
vendor/bin/sail up
```

Now, when you edit the code (in the PHP files of your Laravel application), the workers will be automatically reloaded. In the output messages of Octane, you will see the following:

```
INFO  Application change detected. Restarting workers...
```

Now that we have the auto-reload functionality, we can explore the Swoole features.

Exploring Swoole features

Swoole has a lot of features that we can use in the Laravel Octane application in order to improve the performance and speed of our application. In this chapter, we are going to explore these functionalities and then in the subsequent chapters, we will use these functionalities. The Swoole functionalities that we are going to explore are as follows:

- Concurrent tasks
- Interval command execution
- Caching
- Tables
- Metrics

Concurrent tasks

With Swoole, it is possible to execute multiple tasks in parallel. To demonstrate this, we are going to implement two functions whose execution takes some time.

To simulate the fact that these two functions are time-consuming, we will use the `sleep()` function, which suspends execution for a certain number of seconds.

These two functions return strings: the first one returns "`Hello`" and the second one returns "`World`".

We are going to set the execution time to 2 seconds, via the `sleep()` function.

In a classic scenario of sequential execution of the two functions, the total time taken would be 4 seconds plus a millesimal due to overheads.

We are going to track the execution time using the `hrtime()` function.

> **Note**
>
> When you need to track the execution time of a series of instructions, the use of `hrtime()` is recommended since it is a function that returns monotonic timestamps. A monotonic timestamp is a time calculated based on a reference point (thus relative) and is not affected by system date changes such as automatic clock adjustments (NTP or Daylight Savings Time updates).

We are also going to use two anonymous functions because this will come in handy in the second example (the example with concurrent execution) to be able to make an easier comparison.

Another consideration, before we take a look at the code, we are going to implement the examples directly in the `routes/web.php` file for simplicity and focus on the example code. You can use that code, especially `Octane::concurrently()` in your controllers or other parts of your Laravel application.

> **Note**
>
> For access to these examples, we are going to use the configuration with Laravel Sail and Swoole, with APP_PORT set to 81. If you are going to use your local Open Swoole installation, refer to 127.0.0.1:8000 instead of 127.0.0.1:81.

The example shows the sequential execution:

```
Route::get('/serial-task', function () {
    $start = hrtime(true);
    [$fn1, $fn2] = [
        function () {
            sleep(2);
            return 'Hello';
        },
        function () {
            sleep(2);
            return 'World';
        },
    ];
    $result1 = $fn1();
    $result2 = $fn2();
    $end = hrtime(true);

    return "{$result1} {$result2} in ".($end - $start) /
        1000000000 .' seconds';
});
```

If you access the http://127.0.0.1:81/serial-taskpage with your browser, you should see this output on your page:

Hello World in 4.001601125 seconds

The two functions are executed sequentially, which means that the execution time is the sum of the execution time of the first function and that of the second function.

If you call the two functions with the Octane::concurrently() method (passing your functions as an array of Closure), you can execute the functions in parallel:

```
use Laravel\Octane\Facades\Octane;
Route::get('/concurrent-task', function () {
```

```
    $start = hrtime(true);
    [$result1, $result2] = Octane::concurrently([
        function () {
            sleep(2);
            return 'Hello';
        },
        function () {
            sleep(2);
            return 'World';
        },
    ]);
    $end = hrtime(true);
    return "{$result1} {$result2} in ".($end - $start) /
      1000000000 .' seconds';
});
```

If you open your browser to http://127.0.0.1:81/concurrent-task, you will see the following message:

Hello World in 2.035140709 seconds

Another thing to note is that for concurrent functions, the execution time depends on what happens inside the function. For example, if you want to execute two or more functions in parallel that take an unpredictable amount of time because they depend on third-party factors such as the response time of web services or the workload of a database, or when a huge file is being parsed, you probably can't make any assumptions about the order of the execution or the duration time.

In the next example, we have two simple functions: the first one takes more time to be executed, so even though it is the first function, it is completed after the second one. This is obvious for people who are used to working with parallel tasks but probably is less obvious for people who are used to using a strong synchronous language (such as PHP without Swoole or other tools that add asynchronous functionalities):

```
use Laravel\Octane\Facades\Octane;
Route::get('/who-is-the-first', function () {
    $start = hrtime(true);
    [$result1, $result2] = Octane::concurrently([
        function () {
            sleep(2);
            Log::info('Concurrent function: First');
```

```
            return 'Hello';
        },
        function () {
            sleep(1);
            Log::info('Concurrent function: Second');
            return 'World';
        },
    ]);
    $end = hrtime(true);
    return "{$result1} {$result2} in ".($end - $start) /
      1000000000 .' seconds';
});
```

The result is to print the second statement in the log file before the first one. If you take a look at the storage/logs/laravel.log file, you will see the following:

```
local.INFO: Concurrent function: Second
local.INFO: Concurrent function: First
```

It means that the execution of the functions called by Octane::concurrently start more or less at the same time, but the exact moment they are completed, the execution, depends on the time needed by the execution of the function.

Why is this important to keep in mind?

Because everything might be okay if the two functions are totally independent of each other. On the other hand, if the functions operate using the same resources (reading and writing to the same database table, for example), we need to consider the dependencies between the two operations. For example, one function might alter the data in the table, while the other function might read it. The time at which the data is read is relevant: think about whether the data is read before it is written or whether it is read after it is written. In this case, we might have two totally different behaviors.

Regardless, we will go more deeply into the concurrently method in the next chapter, where we will use Octane in a more real-life scenario – for example, using concurrently to retrieve data from multiple database queries and multiple API calls.

Interval command execution

Sometimes, you have to execute a function every X second(s). For example, you want to execute a function every 10 seconds. With Swoole, you can use the Octane::tick() function, where you can provide a name (as the first parameter) and the function defining a Closure as the second parameter.

The best place to call the `tick()` function is in the `boot()` method of one of your service providers. Typically, I use the default `AppServiceProvider` in the `app/Providers` directory, already created for you when you set up a new Laravel application (with the `laravel new` command, for example).

In `app/Providers/AppServiceProvider.php`, in the `boot()` method, call the `Octane::tick()` function with a very simple feature that logs a message with the timestamp. The log message will be tracked in the `storage/logs/laravel.log` file unless you have some special configuration in the `.env` file:

```
public function boot()
{
    Octane::tick('simple-ticker', fn () =>
    Log::info('OCTANE TICK.', ['timestamp' => now()]))
    ->seconds(10)
    ->immediate();
}
```

In the preceding code snippet, we are using `Octane` and `Log` classes, so remember to include them at the top of the `AppServiceProvider.php` file:

```
use Laravel\Octane\Facades\Octane;
use Illuminate\Support\Facades\Log;
```

The `Octane::tick()` method returns the `InvokeTickCallable` object that implements a few methods:

- `__invoke()`: A special method used for invoking the `tick()` listener; it is the method responsible for executing the function passed as the second parameter to the `tick()` method.

- `seconds()`: A method for indicating how often the listener should be automatically invoked (via the `__invoke()` method). It accepts an integer parameter in seconds.

- `immediate()`: A method indicating that the listener should be invoked on the first tick (so, as soon as possible).

Note

The `tick()` method from the application service is called when you start the `octane:start` command. If you are using Laravel Sail, the application service is loaded and booted when you run `sail up`, just because at the end, Sail launches `supervisord` and the configuration of `supervisord` has the `octane:start` command.

Once Octane is started in `storage/logs/laravel.log`, you can see the message logged by the `tick()` function:

```
[2022-07-29 08:23:11] local.INFO: OCTANE TICK.
{"timestamp":"2022-07-29 08:23:11"}
[2022-07-29 08:23:22] local.INFO: OCTANE TICK.
{"timestamp":"2022-07-29 08:23:21"}
[2022-07-29 08:23:31] local.INFO: OCTANE TICK.
{"timestamp":"2022-07-29 08:23:31"} [2022-07-26 20:45:19]
local.INFO: OCTANE TICK. {"timestamp":"2022-07-26 20:45:19"}
```

In the snippet here, the Laravel log shows us the execution of the `tick()` method.

> **Note**
>
> For displaying the log in real time, you can use the `tail -f` command – for example, `tail -f storage/logs/laravel.log`.

Caching

Managing a caching mechanism to momentarily save some data in a system based on workers, where each worker has its own memory space, might not be so straightforward.

Laravel Octane provides a mechanism to non-permanently save data shared between different workers. This means that if a worker needs to store a value and make it available to subsequent executions by other workers, it is possible to do so through the cache mechanism. The caching mechanism in Laravel Octane is implemented through Swoole Table, which we will see in detail later.

In this specific case, we are going to use the `Cache` class exposed directly by Laravel. Laravel's caching mechanism allows us to use different drivers, such as, for example, databases (MySQL, Postgresql, or SQLite), Memcache, Redis, or other drivers. Because we installed Laravel Octane and used Swoole, we can use a new Octane-specific driver. If we wanted to use Laravel's caching mechanism, we would use the `Cache` class with the classic `store` method to store a value. In the example, we are going to store a key called `last-random-number` in the Octane driver, to which we are going to associate a random number. In the example, we're going to call the function responsible to store the value in the cache via the Octane tick seen earlier with an interval set to 10 seconds. We will see that every 10 seconds, a new cached value is generated with a random number. We are also going to implement a new route, `/get-number`, where we are going to read this value and display it on a web page.

To obtain the Cache instance with the Octane provider you can use `Cache::store('octane')`. Once you have the instance, you can use the `put()` method to store the new value.

In the app/Providers/AppServiceProvider.php file, in the boot() method, add the following:

```
Octane::tick('cache-last-random-number',
    function () {
        $number = rand(1, 1000);
        Cache::store('octane')->put(
            'last-random-number', $number);
        Log::info("New number in cache: ${number}",
            ['timestamp' => now()]);
        return;
    }
)
->seconds(10)
->immediate();
```

Be sure that you include the right classes at the beginning of your file. The classes needed by this code snippet are as follows:

```
use Illuminate\Support\Facades\Cache;
use Laravel\Octane\Facades\Octane;
use Illuminate\Support\Facades\Log;
```

If you want to check, you can see the log messages printed by the tick() function in the storage/logs/laravel.log file.

Now, we can create a new route in the routes/web.php file, where we get the value (last-random-number) from the cache:

```
use Illuminate\Support\Facades\Cache;
Route::get('/get-random-number', function () {
    $number = Cache::store('octane')->get(
        'last-random-number', 0);
    return $number;
});
```

If you open your browser and you go to your /get-random-number path (in my case, http://127.0.0.1:81/get-random-number because I'm using Laravel Sail and I set APP_PORT=81 in the .env file), you can see the random number. If you refresh the page, you will see the same number for 10 seconds, or in general for the interval time set with the tick() function (in the service provider file).

> **Note**
>
> The storage in the Octane cache is not permanent; it is volatile. This means that you can lose the values every time the server is restarted.

With the Octane cache, we have also some other nice methods, such as increment and decrement, for example:

```
Route::get('/increment-number', function () {
    $number =
      Cache::store('octane')->increment('my-number');
    return $number;
});
Route::get('/decrement-number', function () {
    $number =
      Cache::store('octane')->decrement('my-number');
    return $number;
});
Route::get('/get-number', function () {
    $number = Cache::store('octane')->get('my-number', 0);
    return $number;
});
```

Now, open your browser and the /increment-number path load multiple times and then load /get-number; you will see the incremented value.

Other useful functions for managing multiple values are putMany(), which saves an array of items, and many(), which retrieves more items at a time:

```
Route::get('/save-many', function () {
    Cache::store('octane')->putMany([
        'my-number' => 42,
        'my-string' => 'Hello World!',
        'my-array' => ['Kiwi', 'Strawberry', 'Lemon'],
    ]);
    return "Items saved!";
});
Route::get('/get-many', function () {
    $array = Cache::store('octane')->many([
        'my-number',
```

```
        'my-string',
        'my-array',
    ]);
    return $array;
});
```

If you open your browser first at the /save-many path and then at /get-many, you will see the following on the page:

```
{"my-number":42,"my-string":"Hello World!","my-array":["Kiwi","
Strawberry","Lemon"]}
```

If you saved a value as an item of an array with the putMany() method, you can retrieve it as a single item using the array key as the cache key:

```
Route::get('/get-one-from-many/{key?}', function ($key = "my-
number") {
    return Cache::store('octane')->get($key);
});
```

If you open the /get-one-from-many/my-string page, you will see the following:

Hello World!

This means that the value with the "my-string" key is retrieved from the cache.

In the end, we are using the Laravel cached mechanism with Swoole as the backend provider.

Why should we use Swoole as a backend cache provider? What advantages does it have over other providers such as databases, for example? The way Swoole is made, we have seen that performance is certainly one of its main values. Using this kind of cache allows us to share information between workers in a very efficient way. Swoole's official documentation reports that it supports a read-write rate of 2 million per second.

The Laravel Octane cache is implemented via Swoole Table. In the next section, we will look at Swoole Table.

Swoole Table

If you want to share data across workers in a more structured way than a cache, you can use Swoole Table. As already mentioned, the Octane cache uses Swoole Table, so let me explain a bit more about Swoole Table.

First, if you want to use Swoole Table, you must define the structure of the table. The structure is defined in the `config/octane.php` file in the `tables` section.

```
/*
|--------------------------------------------------------------------------
| Octane Swoole Tables
|--------------------------------------------------------------------------
|
| While using Swoole, you may define additional tables as required by the
| application. These tables can be used to store data that needs to be
| quickly accessed by other workers on the particular Swoole server.
|
*/

'tables' => [
    'example:1000' => [
        'name' => 'string:1000',
        'votes' => 'int',
    ],
],
```

Figure 3.5 – The configuration of a table in a config/octane.php file

By default, when you execute the `octane:install` command, a `config/octane.php` file is created for you, and a table is already defined: a table named `example` with a maximum of 1,000 rows (`example:1000`) with 2 fields. The first one is called `name` and is a string with a maximum of 500 characters, and the second field is called `votes` and the type is an integer (`int`).

In the same way, you can configure your table. The field types are as follows:

- `int`: For integers
- `float`: For floating-point numbers
- `string`: For strings; you can define the max number of characters

Once you have defined the table in the config file, you can set the values of the table.

For example, I'm going to create a `my-table` table of max 100 rows with some fields. In the `config/octane.php` file in the `tables` section, we are going to create the following:

- A table named `my-table` with a maximum of 100 rows
- A UUID field with a `string` type (maximum of 36 characters)
- A name field with a `string` type (maximum of 1000 characters)
- An age field with an `int` type
- A value field with a `float` type

Therefore, the configuration is as follows:

```
'my-table:100' => [
    'uuid' => 'string:36',
    'name' => 'string:1000',
    'age' => 'int',
    'value' => 'float',
],
```

When the server starts, Octane will create the table in memory for us.

Now, we are going to create two routes: the first one is for seeding the table with some *fake* data, and the second one is for retrieving and showing information from the table.

The first route, /table-create, will do the following:

- Obtain the table instance for my-table. my-table is the table configured in config/octane.php.

- Create 90 rows, filling fields for uuid, name, age, and value. For filling the values, we are going to use the fake() helper. This helper allows you to generate random values for UUIDs, names, integer numbers, decimal numbers, and so on.

The code in routes/web.php is as follows:

```
Route::get('/table-create', function () {
    // Getting the table instance
    $table = Octane::table('my-table');
    // looping 1..90 creating rows with fake() helper
    for ($i=1; $i <= 90; $i++) {
        $table->set($i,
        [
            'uuid' => fake()->uuid(),
            'name' => fake()->name(),
            'age' => fake()->numberBetween(18, 99),
            'value' => fake()->randomFloat(2, 0, 1000)
        ]);
    }
    return "Table created!";
});
```

The snippet creates the table once the /table-create URL is called.

If you open http://127.0.0.1:81/table-create, you will see a table is created with some rows.

On your page, it's possible that you have an error such as this one:

```
Swoole\Table::set(): failed to set('91'), unable to allocate
memory
```

Be sure that the size of the table in the configuration file is greater than the number of rows that you are creating. When we work with a database table, we don't have to set the maximum number of rows; in this case, with this kind of table (in a memory table shared across the workers), we have to be aware of the number of rows.

Once everything is fine, you will see **Table created!** on your web page. This means that the rows were created in the right way.

To verify this, we are going to create a new route, /table-get, where we do the following:

- Obtain the table instance for my-table (the same table we used in /table-create)
- Get the rows with index 1
- Return the row (associative array, where the items are the fields of the row with uuid, name, age, and value)

In the routes/web.php file, define the new route:

```
Route::get('/table-get', function () {
    $table = Octane::table('my-table');
    $row = $table->get(1);
    return $row;
});
```

Opening https://127.0.0.1:81/table-get (after opening /table-create because you have to create rows before accessing them), you should see something such as the following:

```
{"uuid":"6e7c7eb6-9ecf-3cf8-8de9-5034f4c44ab5","name":"Hugh
Larson IV","age":81,"value":945.67}
```

You will see something different in terms of content because rows were generated using the fake() helper, but you will see something similar in terms of structure.

You can also walk through Swoole Table using `foreach()`, and you can also use the `count()` function (for counting elements of a table) on the `Table` object:

```
Route::get('/table-get-all', function () {
    $table = Octane::table('my-table');
    $rows=[];
    foreach ($table as $key => $value) {
        $rows[$key] = $table->get($key);

    }
    // adding as first row the table rows count
    $rows[0] = count($table);
    return $rows;
});
```

In the preceding example, for counting rows, you can see that in the end, we can access the Swoole object and use the functions and methods implemented by it. Another example of using a Swoole object is accessing the server object, as in the next section for retrieving metrics.

Metrics

Swoole offers a method to retrieve some metrics about the resource usage of the application server and the workers. This is useful if you want to track the usage of some aspect – for example, the time, the number of requests served, the active connections, the memory usage, the number of tasks, and so on. To retrieve metrics, first of all, you have to access the Swoole class instance. Thanks to the service container provided by Laravel, you can resolve and access objects from the container. The Swoole server object is *stored* in the container by Octane so, via `App::make`, you can access the Swoole server class instance. In the `routes/web.php` file, you can create a new route where you can do the following:

- You can retrieve the Swoole server via `App::make`

- Once you can access the object, you can use their methods, such as, for example, `stats`

- The Swoole server `Stats` object is returned as a response:

```
use Illuminate\Support\Facades\App;
Route::get('/metrics', function () {
    $server = App::make(Swoole\Http\Server::class);
    return $server->stats();
});
```

On the new /metrics page, you can see all the metrics provided by the Swoole server.

For all methods, you can see directly the server documentation provided by Swoole: https://openswoole.com/docs/modules/swoole-server-stats.

Summary

In this chapter, we looked at how to install and configure an environment with Swoole. Then, we also analyzed the various functionalities exposed by this application server.

We walked through the functionalities with simple examples so that we could focus on the individual features. In the next chapter, we will look at using Laravel Octane with an application server in a more real-world context.

Part 3:
Laravel Octane –
a Complete Tour

The goal of this part is to show practical examples of how to best use the features provided by Laravel Octane and the Swoole application server, especially for optimized data access. Examples of data query management with caching and parallel queries are shown in this part.

This part comprises the following chapters:

- *Chapter 4, Building a Laravel Octane Application*
- *Chapter 5, Reducing Latency and Managing Data with an Asynchronous Approach*

4

Building a Laravel Octane Application

In the previous chapters, we focused on installing, configuring, and using some of the features provided by Laravel Octane. We looked at the difference between Swoole and RoadRunner, the two application servers supported by Laravel Octane.

In this chapter, we will focus on each feature of Laravel Octane to discover its potential and understand how it can be used individually.

The goal of this chapter is to analyze the functionality of Laravel Octane in a realistic context.

To do that, we will build a sample dashboard application covering several aspects, such as configuring the application, creating the database table schema, and generating initial data.

Then, we will go on to implement specific routes for dashboard delivery and implement the data retrieval logic in the controller and the query in the model.

Then we will create a page in the sample dashboard application, where we will collect information from multiple queries. When we have to implement queries for retrieving data, generally, we focus on the logic and the methods for filtering, sorting, and selecting data. In this chapter, however, we will keep the logic as simple as possible to allow you to focus on other aspects, such as loading data efficiently thanks to executing parallel tasks, and we will apply some strategies to reduce the response time as much as possible (running the tasks in parallel reduces the overall response execution time).

In designing the application architecture, we also need to consider the things that could go wrong.

In the examples in the previous chapter, we analyzed each feature by considering what is called the **happy path**. The happy path is the default scenario that the user takes to achieve the desired result without encountering any errors.

In designing a real application, we must also think about all those cases that are not included in the happy path. For example, in the case of concurrent execution of heavy queries, we need to think about the case where the execution may return an unexpected result such as an empty result set, or when the execution of a query raises an exception. We need to consider that this single exception may also have an impact on other concurrent executions. This looks like a more real-life scenario (where things could go wrong because of some exceptions) and, in this chapter, we will try to also manage the errors.

Therefore, we will try to simulate a typical data-consuming application, where users' request response controllers must execute operations as fast as possible, even in the face of a high request load.

The primary objective of this chapter is to guide you through drastically reducing the response time of your application with the help of multiple queries, concurrent execution in rendering a dashboard page, and by trying to apply Octane features in the application. We will walk through the routing, controller, models, queries, migrations, seeding, and the view template. We will involve some mechanisms provided by Octane, such as Octane Routes, chunk data loads, parallel tasks (for queries and HTTP requests), error and exception management, and Octane Cache.

In this chapter, we will cover the following:

- Installing and setting up the application
- Importing the initial data (and suggestions on how to make it efficiently)
- Querying multiple pieces of data from the database in parallel
- Optimizing the routes
- More examples of integrating third-party APIs
- Improving speed with Octane Cache

Technical requirements

We are going to assume that you have PHP 8.0 or greater (8.1 or 8.2) and the Composer tool. If you want to use Laravel Sail (`https://laravel.com/docs/9.x/sail`), you need the Docker Desktop application (`https://www.docker.com/products/docker-desktop`).

We will also quickly recap the setup of Octane for our practical example. So, we will install all tools needed.

The source code and the configuration files of the examples described in the current chapter are available here: `https://github.com/PacktPublishing/High-Performance-with-Laravel-Octane/tree/main/octane-ch04`

Installing and setting up the dashboard application

To demonstrate the power of Laravel Octane, we are going to build a dashboard page to show event data filtered in different ways. We will keep it as simple as possible to avoid focusing on business functionalities, and we will keep focusing on how to apply techniques for improving performance while keeping the application reliable and error-free.

Installing your Laravel application

As shown in *Chapter 1, Understanding the Laravel Web Application Architecture*, you can install the Laravel application from scratch via the Laravel command, as follows:

```
composer global requires laravel/installer
```

Once you have your Laravel command installed, you can create your application with the following command:

```
laravel new octane-ch04
```

The `laravel new` command creates the directory with your application, so the next step is to enter the new directory to start customizing the application:

```
cd octane-ch04
```

Adding a database

Now that we have created the application, we have to install and set up the database because our example application will need a database to store and retrieve the example data. So, to install and set up the database, we are going to do the following:

1. Install a MySQL database server
2. Execute migrations in Laravel (to apply schema definitions to database tables)
3. Install an application to manage and check the tables and the data of the database

Installing the database service

There are three ways to install the database server: via the official installer, via your local package manager, or via Docker/Laravel Sail.

The first one is to use the official installer provided by MySQL. You can download and execute the installer from the official website for your specific operating system: https://dev.mysql.com/downloads/installer/.

Once you have downloaded the installer, you can execute it.

Another way is to use your system package manager. If you have macOS, my suggestion is to use Homebrew (see *Chapter 1, Understanding the Laravel Web Application Architecture*) and execute the following command:

```
brew install mysql
```

If you are using GNU/Linux, you can use the package manager provided by your GNU/Linux distribution. For example, for Ubuntu, you can execute the following:

```
sudo apt install mysql-server
```

If you don't want to install or add the MySQL server to your local operating system, you can use a Docker image running in a Docker container. For that, we can use the **Laravel Sail** tool. If you are familiar with Docker images, using a Docker image simplifies the installation of third-party software (such as the database). Laravel Sail simplifies the process of managing Docker images.

Make sure that Laravel Sail is added to your application. In the project directory, add the Laravel Sail package to your project:

```
composer require laravel/sail --dev
```

Then, execute the new command provided by Laravel Sail to add the Sail configuration for Docker:

```
php artisan sail:install
```

The execution of the preceding command will require you to select the services you need to activate via Laravel Sail. For now, the goal is to activate the MySQL service, so select the first option. On selecting the MySQL service, the MySQL Docker image will automatically be downloaded:

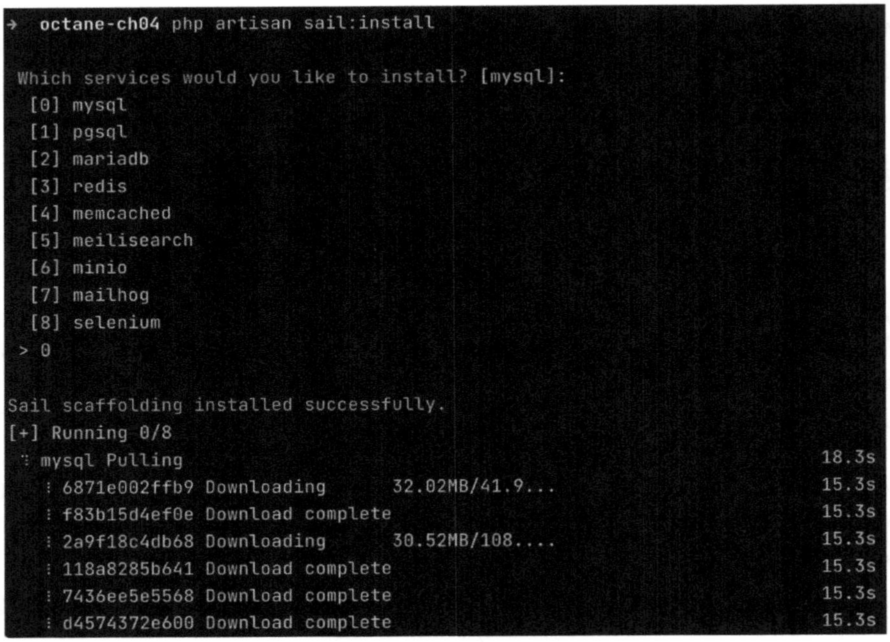

Figure 4.1: Installing Laravel Sail

Installing Laravel Sail, as well as downloading the MySQL Docker image, will add the `docker-compose.yml` file to your project directory, and the PHPUnit configuration will be changed to use the new database instance. So, installing Laravel Sail helps you with the Docker configuration (creating the `docker-compose.yml` file with a preset configuration based on the choices provided as answers to the questions raised by the `sail:install` command), and with the configuration of PHPUnit (creating the right PHPUnit configuration to use the new database instance).

The `docker-compose.yml` file will contain the following:

- The main service to serve your web application
- An additional service for the MySQL server
- The right configuration for the services to use the same environment variables from the `.env` file

If you already have some services up and running on your local operating system and you want to avoid some conflicts (multiple services that use the same port), you can control some parameters used by Docker containers via `docker-compose.yml`, setting the following variables in the `.env` file:

- `VITE_PORT`: This is the port used by Vite to serve the frontend part (JavaScript and CSS). The default is `5173`; if you have Vite already up and running locally, you could use port `5174` to avoid conflicts.
- `APP_PORT`: This is the port used by the web server. By default, the port used by the local web server is port `80`, but if you already have a local web server up and running, you can use the `8080` settings (`APP_PORT=8080`) in the `.env` file.
- `FORWARD_DB_PORT`: This is the port used by Laravel Sail to expose the MySQL service. By default, the port used by MySQL is `3306`, but if it is already in use, you can set the port via `FORWARD_DB_PORT=3307`.

Once the `.env` configuration is good for you, you can start the Docker containers via Laravel Sail.

To start Laravel Sail and launch the Docker container, use the following command:

```
./vendor/bin/sail up -d
```

The `-d` option allows you to execute Laravel Sail in the background, which is useful if you want to reuse the shell to launch other commands.

To check that your database is up and running, you can execute the `php artisan db:show` command via `sail`:

```
./vendor/bin/sail php artisan db:show
```

The first time you execute the db:show command, an additional package – the Doctrine **Database Abstraction Layer** (**DBAL**) package – will be installed automatically in your Composer dependencies. The Doctrine DBAL package will add database inspection functionalities to the artisan command. Once you run the db:show command, this is what you'll see:

```
→  octane-ch04 git:(main) ✗ ./vendor/bin/sail php artisan db:show

MySQL 8 .....................................................................
Database ........................................................ octane_ch04
Host ................................................................... mysql
Port .................................................................... 3306
Username ................................................................ sail
URL ........................................................................
Open Connections ........................................................... 1
Tables ..................................................................... 0
```

Figure 4.2: Executing the db:show command via Sail

Now your database is up and running, so you can create your tables. We are going to execute migration to create the database tables. The database tables will contain your data – for example, the events.

A migration file is a file where you can define the structure of your database table. In the migration file, you can list the columns of your table and define the type of the columns (string, integer, date, time, etc.).

Executing the migration

The Laravel framework provides out-of-the-box migrations specific to standard functionalities such as user and credential management. That's why after installing the framework in the database/ migrations directory you can find migration files already provided with the framework: the migrations to create a users table, a password resets table, a failed jobs table, and a personal access tokens table.

The migration files are stored in the database/migrations directory.

To execute the migration in the Docker container, you can execute the migrate command via the command line:

```
./vendor/bin/sail php artisan migrate
```

This is what you'll see:

```
octane-ch04 git:(main) × ./vendor/bin/sail php artisan migrate

   INFO   Preparing database.

  Creating migration table ........................................ 45ms DONE

   INFO   Running migrations.

  2014_10_12_000000_create_users_table ............................ 49ms DONE
  2014_10_12_100000_create_password_resets_table .................. 29ms DONE
  2019_08_19_000000_create_failed_jobs_table ...................... 27ms DONE
  2019_12_14_000001_create_personal_access_tokens_table ........... 41ms DONE
```

Figure 4.3: Executing migrations

If you are not using Laravel Sail, and you are using the MySQL server installed in your local operating system (with Homebrew or your operating system packager or the MySQL server official installer), you can use `php artisan migrate` without the `sail` command:

```
php artisan migrate
```

The schema of the database and the tables are created thanks to the migrations. Now we can install the MySQL client to access the database.

Installing MySQL client

To access the structure and the data of the database, it is recommended that you install a MySQL client. The MySQL client allows you to access the structure, the schema, and the data and allows you to execute SQL queries to extract data.

You can choose one of the tools available; some are open source, and others are paid tools. The following shows some of the tools for managing MySQL structures and data:

- **Sequel Ace** is open source, and is available for macOS: `https://github.com/Sequel-Ace/Sequel-Ace`

- **MySQL Workbench** is the official one and is available for all platforms: `https://www.mysql.com/products/workbench/`

- **TablePlus** is available for Windows and macOS and supports a lot of databases: `https://tableplus.com/`

If you select Sequel Ace or other tools, you have to set the right parameters during the initial connection, according to the .env file.

For example, the initial screen of Sequel Ace asks you for the hostname, the credential, the database name, and the port:

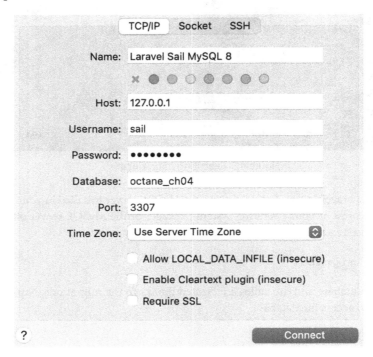

Figure 4.4: The Sequel Ace login screen

As shown in *Figure 4.4*, here are the values:

- **Host**: 127.0.0.1
- **Username**: The DB_USERNAME parameter in the .env file
- **Password**: The DB_PASSWORD parameter in the .env file
- **Database**: The DB_DATABASE parameter in the .env file
- **Port**: The FORWARD_DB_PORT parameter if you are using Laravel Sail, or DB_PORT if you are not using a local Docker container

After installing the MySQL client, we'll move on to talking about Sail versus the local tools.

Sail versus local tools

We looked at two methods for using PHP, services, and tools: using Docker containers (Laravel Sail) and using a local installation.

Once Sail is set up, if you want to launch commands via Sail, you have to prefix your command with `./vendor/bin/sail`. For example, if you want to list the PHP modules that are installed, the following command will list all PHP modules installed on your local operating system:

```
php -m
```

If you use the `php -m` command with the `sail` tool, as shown in the following, the PHP modules installed in the Docker container will be shown:

```
./vendor/bin/sail php -m
```

The Laravel Sail image provides you with the Swoole extension already installed and configured, so now you can add Octane to your application.

Adding Octane to your application

To add Laravel Octane to your application, you have to do the following:

1. Add the Octane package
2. Create Octane configuration files

Information

We already covered the Octane setup with Laravel Sail and Swoole in *Chapter 3*, *Using the Swoole Application Server*. Let's quickly recap all the steps for the Octane configuration needed by the example provided to you in the current chapter.

So, first of all, in the project directory, we are going to add the Laravel Octane package with the `composer require` command:

```
./vendor/bin/sail composer require laravel/octane
```

Then, we will create Octane configuration files with the `octane:install` command:

```
./vendor/bin/sail php artisan octane:install
```

Now that we have installed Laravel Octane, we have to configure Laravel to start the Swoole application server.

Activating Swoole as the application server

If you are using Laravel Sail, you have to activate Swoole to serve your Laravel application. The default Laravel Sail configuration launches the classical php artisan serve tool. So, the goal is to edit the configuration file where the artisan serve command is defined and replace it with the octane:start command. To do that, you have to copy the configuration file from the vendor directory to a directory where you can edit it. Laravel Sail provides you a publishing command to copy and generate the configuration file via the sail:publish command:

./vendor/bin/sail artisan sail:publish

The publish command generates the Docker directory and the supervisord.conf file. The supervisord.conf file has the responsibility of launching the web service to accept the HTTP request and generate the HTTP response. With Laravel Sail, the command that runs the web service is placed in the supervisord.conf file. Then, in the docker/8.1/supervisord.conf file (placed in the project directory), to launch Laravel Octane instead of the classical web server, replace the artisan serve command with artisan octane:start with all the correct parameters:

```
# command=/usr/bin/php -d variables_order=EGPCS /var/www/html/
artisan serve --host=0.0.0.0 --port=80
command=/usr/bin/php -d variables_order=EGPCS /var/www/html/
artisan octane:start --server=swoole --host=0.0.0.0 --port=80
```

With Laravel Sail, when you change any Docker configuration files, you must rebuild the images:

./vendor/bin/sail build --no-cache

Then, restart Laravel Sail:

./vendor/bin/sail stop
./vendor/bin/sail up -d

If you open your browser to http://127.0.0.1:8080/, you will see your Laravel application served by Swoole.

Verifying your configuration

Once you set up the tools and services, my suggestion is to be aware of the configuration used by the tools. With the PHP command, you have some options to check the installed module (useful to check whether a module is loaded correctly, for example, to check whether the Swoole module is loaded), and an option to see the current configuration of PHP.

To check whether a module is installed or not, you can use the PHP command with the -m option:

```
./vendor/bin/sail php -m
```

To check whether Swoole is correctly loaded, you can filter just the lines with Swoole as the name (case-insensitive). To filter the lines, you can use the grep command. The grep command shows only the lines that match specific criteria:

```
./vendor/bin/sail php -m | grep -i swoole
```

If you want to list all the PHP configurations, you can use the -i option:

```
./vendor/bin/sail php -i
```

If you want to change something in your configuration, you might want to see where the configuration (.ini) files are located. To see where the .ini files are located, filter just the ini string:

```
./vendor/bin/sail php -i | grep ini
```

You will see something like this:

```
Configuration File (php.ini) Path => /etc/php/8.1/cli
Loaded Configuration File => /etc/php/8.1/cli/php.ini
Scan this dir for additional .ini files => /etc/php/8.1/cli/
conf.d
Additional .ini files parsed => /etc/php/8.1/cli/conf.d/10-
mysqlnd.ini,
```

With the php -i command, you can obtain information about where the php.ini file is located. If you are using Laravel Sail, you can execute the following command:

```
./vendor/bin/sail php -i | grep ini
```

You will see that there is a specific .ini file for Swoole:

```
/etc/php/8.1/cli/conf.d/25-swoole.ini
```

If you want to access that file to check it or edit it, you can jump into the running container via the shell command:

```
./vendor/bin/sail shell
```

With this command, it will show the shell prompt of the running container, and you can show the content of the file there:

```
less /etc/php/8.1/cli/conf.d/25-swoole.ini
```

The command will show you the content of the 25-swoole.ini configuration file. The content of the file is as follows:

```
extension=swoole.so
```

If you want to disable Swoole, you can add the ; character at the beginning of the extension directive, as follows:

```
; extension=swoole.so
```

With the ; character at the beginning, the extension is not loaded.

Summarizing installation and setup

Before proceeding with implementation, let me summarize the previous steps:

1. We installed our Laravel application.
2. We added a database service.
3. We configured a MySQL client to access the MySQL server.
4. We added the Octane package and configuration.
5. We added Swoole as the application server.
6. We checked the configuration.

So, now we can start using some Octane functionalities such as executing heavy tasks in a parallel and async way.

Creating a dashboard application

In an application, you can have multiple kinds of data stored in multiple tables.

Typically, on the product list page, you have to retrieve a list of products by executing a query to retrieve products.

Or, in a dashboard, maybe you could show multiple charts or tables to show some data from your database. If you want to show more charts on the same page, you have to perform more than one query on more than one table.

You might execute one query at a time; this means that the total time for retrieving all the useful information for composing the dashboard is the sum of the execution times of all the queries involved.

Running more than one query at the same time would reduce the total time to retrieve all the information.

To demonstrate this, we will create an events table where we will store some events with a timestamp for the user.

Creating an events table

When you are creating a table in Laravel, you have to use a migration file. A migration file contains the logic to create the table and all fields. It contains all the instructions to define the structure of your table. To manage the logic for using the data stored in the table, you might need other things such as the model and seeder classes.

The model class allows the developer to access the data and provides some methods for saving, deleting, loading, and querying data.

The seeder class is used to fill the table with initial values or sample values.

To create the model class, the seeder class, and the migration file, you can use the make:model command with the m (create a migration file) and s (create a seeder class) parameters:

```
php artisan make:model Event -ms
```

With the make:model command and the m and s parameters, three files are created:

- The migration file is created in database/migration/, with the name consisting of the timestamp as the prefix and create_events_table as the suffix, for example, 2022_08_22_210043_create_events_table.php

- The model class in app/Models/Event.php

- The seeder class file in app/database/seeders/EventSeeder.php

Customizing the migration file

The make:model command creates a template file for creating the table with basic fields such as id and timestamps. The developer must add the fields specific to the application. In the dashboard application, we are going to add these fields:

- user_id: For the external reference with the users table, a user could be related to more events

- type: An event type could be INFO, WARNING, or ALERT

- description: Text containing the description of the event

- `value`: An integer from 1 to 10
- `date`: The event date and time

To create the table, an example of the migration file is as follows:

```php
<?php

use App\Models\User;
use Illuminate\Database\Migrations\Migration;
use Illuminate\Database\Schema\Blueprint;
use Illuminate\Support\Facades\Schema;

return new class extends Migration
{
    /**
     * Run the migrations.
     *
     * @return void
     */
    public function up()
    {
        Schema::create('events',
                    function (Blueprint $table) {
            $table->id();
            $table->foreignIdFor(User::class)->index();
            $table->string('type', 30);
            $table->string('description', 250);
            $table->integer('value');
            $table->dateTime('date');
            $table->timestamps();
        });
    }

    /**
     * Reverse the migrations.
     *
     * @return void
```

```
    */
    public function down()
    {
        Schema::dropIfExists('events');
    }
};
```

You can list the fields you want to add to the table in the up() method. In the code, we are adding the foreign ID for the user table, the type, the description, the value, and the date. The down() method typically is used to drop the table. The up() method is called when the developer wants to execute the migrations, and the down() method is called when the developer wants to roll back the migration.

Seeding data

With the seeder file, you can create the initial data to fill the table. For testing purposes, you can fill the table with fake data. Laravel provides you with a great helper, fake(), for creating fake data.

> **The fake() helper**
> For generating fake data, the fake() helper uses the **Faker** library. The home page of the library is at https://fakerphp.github.io/.

Now, we are going to create fake data for users and events.

To create fake users, you can create the app/database/seeders/UserSeeder.php file.

In the example, we will do the following:

- Generate a random name via fake()->firstName()

- Generate a random email via fake()->email()

- Generate a random hashed password with Hash::make(fake()->password())

We will generate 1,000 users, so we will use a for loop.

You have to generate data and call User::insert() to generate data in the run() method of the UserSeeder class:

```
<?php

namespace Database\Seeders;

use App\Models\User;
```

```php
use Illuminate\Database\Seeder;
use Illuminate\Support\Facades\Hash;

class UserSeeder extends Seeder
{
    /**
     * Run the database seeds.
     *
     * @return void
     */
    public function run()
    {
        $data = [];
        $passwordEnc = Hash::make(fake()->password());
        for ($i = 0; $i < 1000; $i++) {
            $data[] =
            [
                'name' => fake()->firstName(),
                'email' => fake()->unique()->email(),
                'password' => $passwordEnc,
            ];
        }
        foreach (array_chunk($data, 100) as $chunk) {
            User::insert($chunk);
        }
    }
}
```

With the UserSeeder class, we are going to create 1,000 users. Then, once we have the users in the user table, we are going to create 100,000 events:

```php
<?php

namespace Database\Seeders;

use App\Models\Event;
use Illuminate\Database\Seeder;
```

```
use Illuminate\Support\Arr;

class EventSeeder extends Seeder
{
    /**
     * Run the database seeds.
     *
     * @return void
     */
    public function run()
    {
        $data = [];
        for ($i = 0; $i < 100_000; $i++) {
            $data[] = [
                'user_id' => random_int(1, 1000),
                'type' => Arr::random(
                    [
                        'ALERT', 'WARNING', 'INFO',
                    ]
                ),
                'description' => fake()->realText(),
                'value' => random_int(1, 10),
                'date' => fake()->dateTimeThisYear(),
            ];
        }
        foreach (array_chunk($data, 100) as $chunk) {
            Event::insert($chunk);
        }
    }
}
```

To create fake events, we need to fill the event fields using the fake() helper. The fields filled for the events table are as follows:

- user_id: We will generate a random number from 1 to 1000

- type: We will use the Arr:random() helper from Laravel to select one of these values: 'ALERT', 'WARNING', or 'INFO'

- `description`: A random text from the `fake()` helper

- `value`: A random integer from 1 to 10

- `date`: A date function provided by a `fake()` helper for generating a day from the current year, `dateTimeThisYear()`

Like we did for the `users` table, we are using the chunking approach to try to improve the speed of the execution of the data generator. For large arrays, the chunking approach allows the code to be more performant because it involves dividing the array into chunks and handling the chunks instead of each record individually. This reduces the number of insertions made to the database.

Improving the speed of the seed operation

Generating a lot of data requires thinking about the cost in terms of *time spent* on the operations.

The two most expensive operations used for data seeding (via the `UserSeeder` class) during the creation of the initial user data are as follows:

- `Hash::make()` takes a fraction of a second because it is CPU-intensive. If you repeat this operation multiple times, in the end, it takes seconds to be executed.

- `array_chunk` can help you reduce the number of calls to the `insert()` method. Consider that the `insert()` method can accept an array of items (multiple rows to insert). Using `insert()` with an array as an argument is much faster in the execution than calling `insert()` for every single row. Each `insert()` execution under the hood (at the database level) has to prepare the transaction operation for the insert, insert the row in the table, adjust all indexes and all metadata for the table, and close the transaction. In other words, each `insert()` operation has some overhead time that you have to consider when you want to call it multiple times. That means each `insert()` operation has an additional overhead cost to ensure that the operation is self-consistent. Reducing the number of such operations reduces the total time of the additional operations.

So, in order to improve the performance in data creation (seeding), we can make some assumptions and we can implement these approaches:

- To create multiple users, it is fine to have the same password for all users. We don't have to implement a sign-in process, we just need a list of users.

- We can create an array of users, and then use the chunking approach for inserting chunks of data (for 1,000 users we insert 10 chunks of 100 users each).

So, in the previous snippet of code for creating users, we used these two kinds of optimizations: reducing the number of hash calls and using `array_chunk`.

In some scenarios, you have to insert and load a huge amount of data into the database. In this case, my suggestion is to load data using some specific features provided by the database, instead of trying to optimize your code.

For example, if you have a multitude of data to load and or transfer from another database, in the case of MySQL, there are two tools.

The first option is using the `INTO OUTFILE` option:

```
select * from events INTO OUTFILE '/var/lib/mysql-files/export-events.txt';
```

Before doing that, you have to be sure that MySQL is allowing you to perform this operation.

Because we will export a huge quantity of data in a directory, we have to list this directory as permitted in the MySQL configuration.

In the `my.cnf` file (the configuration file for MySQL), be sure that there is a `secure-file-priv` directive. The value of this directive would be a directory where you can export and import the file.

If you are using Laravel Sail, `secure-file-priv` is already set to a directory:

```
secure-file-priv=/var/lib/mysql-files
```

In the case of Homebrew, the configuration file is located in the following: `/opt/homebrew/etc/my.cnf`.

For example, the `my.cnf` file could have this structure:

```
[mysqld]
bind-address = 127.0.0.1
mysqlx-bind-address = 127.0.0.1
secure-file-priv = "/Users/roberto"
```

In this case, the directory for exporting data and files is `"/Users/roberto"`:

```
opt > homebrew > etc >  ≡ my.cnf
1    # Default Homebrew MySQL server config
2    [mysqld]
3    # Only allow connections from localhost
4    bind-address = 127.0.0.1
5    mysqlx-bind-address = 127.0.0.1
6    secure-file-priv = "/Users/roberto"
7
```

Figure 4.5: The secure-file-priv directive of MySQL

This directive exists for security reasons, so before making this edit, make your evaluation. In the case of the production environment, I disable that directive (set as an empty string). In local development environments, this configuration could be acceptable, or at least activate this option only when you need it.

After this configuration change, you have to reload the MySQL server. In the case of Homebrew, use the following:

```
brew services restart mysql
```

Now you can execute an artisan command (php artisan db) to access the database. You don't need to specify the database name, username, or password because the command uses the Laravel configuration (the DB_ parameters in .env):

```
php artisan db
```

In the MySQL prompt that is shown after you launched the artisan db command, you can, for example, export data using the SELECT syntax:

```
select * from events INTO OUTFILE '/Users/roberto/export-
events.txt';
```

You will see that exporting thousands and thousands of records will take just a few milliseconds.

If you are using Laravel Sail, as usual, you have to launch php artisan through the sail command:

```
./vendor/bin/sail php artisan db
```

In the MySQL Docker prompt use the following:

```
select * from events INTO OUTFILE '/var/lib/mysql-files/export-
events.txt';
```

If you want to load a file that previously exported my SELECT statement, you can use LOAD DATA:

```
LOAD DATA INFILE '/Users/roberto/export-events.txt' INTO TABLE
events;
```

Again, you will see that this command will take a few milliseconds to import thousands and thousands of records:

```
mysql> LOAD DATA INFILE '/Users/roberto/export-events.txt' INTO TABLE events;
Query OK, 100000 rows affected (0.58 sec)
Records: 100000  Deleted: 0  Skipped: 0  Warnings: 0
```

Figure 4.6: With LOAD DATA, you can boost the loading data process

So, in the end, you have more than one way to boost the loading data process. I suggest using LOAD DATA when you have MySQL, and you can obtain data exported via SELECT. Another scenario is when, as a developer, you receive a huge data file from someone else, and you can agree with the file format. Or, if you already know that you will have to load huge amounts of data multiple times for testing purposes, you could evaluate creating a huge file once (for example, with the fake() helper) and then use the file every time you want to seed the MySQL database.

Executing the migrations

Now, before implementing the query to retrieve data, we have to run the migration and the seeders.

So, in the previous sections, we covered how to create seeders and migration files.

To control which seeder has to be loaded and executed, you have to list the seeders in the database/seeders/DatabaseSeeder.php file, in the run() method. You have to list the seeders in this way:

```
$this->call([
    UserSeeder::class,
    EventSeeder::class,
]);
```

To create tables and load data with one command, use this:

```
php artisan migrate --seed
```

If you already executed the migration and you want to recreate them from scratch, you can use migrate:refresh:

```
php artisan migrate:refresh --seed
```

Or you can use the migrate:fresh command, which drops tables instead of executing the rollback:

```
php artisan migrate:fresh --seed
```

> **Note**
> The migrate:refresh command will execute all down() functions of your migrations. Usually, in the down() method, the dropIfExists() method (for dropping the table) is called, so your table will be cleaned and your data will be lost before being created again from scratch.

Now that you have your tables and data created, we will load the data via a query from the controller. Let's see how.

The routing mechanism

As a practical exercise, we want to build a dashboard. A dashboard collects some information from our `events` table. We have to run multiple queries to collect some data to render the dashboard blade view.

In the example, we will do the following:

- Define two routes for `/dashboard` and `/dashboard-concurrent`. The first one is for sequential queries, and the second one is for concurrent queries.

- Define a controller named `DashboardController` with two methods – `index()` (for the sequential queries) and `indexConcurrent()` (for the concurrent queries).

- Define four queries: one for counting the rows in the `events` table, and three queries for retrieving the last five events that include a specific term in the description field (in the example we are looking for the strings that include the term `something`), for each event type (`'INFO'`, `'WARNING'`, and `'ALERT'`).

- Define a view to show the result of the queries.

Using the Octane routes

Octane provides an implementation of a routing mechanism.

The routing mechanism provided by Octane (`Octane::route()`) is lighter than the classic Laravel routing mechanism (`Route::get()`). The Octane routing mechanism is faster because it skips all the full features provided by Laravel routes such as middleware. Middleware is a way of adding functionalities when a route is invoked, but it takes time to call and manage this software layer.

To call Octane routes, you can use the `Octane::route()` method. The `route()` method has three parameters. The first parameter is the HTTP method (for example `'GET'`, `'POST'`, etc.), the second parameter is the path (such as '`/dashboard`'), and the third parameter is a function that returns the `Response` object.

Now that we understand the syntax differences between `Route::get()` and `Octane::route()`, we can modify the last code snippet by replacing `Route::get()` with `Octane::route()`:

```
use Laravel\Octane\Facades\Octane;
use Illuminate\Http\Response;
use App\Http\Controllers\DashboardController;
Octane::route('GET', '/dashboard', function() {
    return new Response(
        (new DashboardController)->index());
});
```

```
Octane::route('GET', '/dashboard-concurrent', function() {
    return new Response(
        (new DashboardController)->indexConcurrent());
});
```

If you want to test how much faster the Octane routing mechanism is than the Laravel routing mechanism, create two routes: the first one served by Octane, and the second one served by the Laravel route. You will see that the response is very fast because the application inherits all the benefits that come from all the Octane framework loader mechanisms, and the Octane::route also optimizes the routing part. The code creates two routes, /a and /b. The /a route is managed via the Octane routing mechanism, and the /b route is managed via the classic routing mechanism:

```
Octane::route('GET', '/a', function () {
    return new Response(view('welcome'));
});
Route::get('/b', function () {
    return new Response(view('welcome'));
});
```

If you compare the two requests by calling it via the browser and checking the response time, you will see that the /a route is faster than the /b route (on my local machine, it is 50% faster) because of Octane::route().

Now that the routes are set up, we can focus on the controller.

Creating the controller

Now we are going to create a controller, DashboardController, with two methods: index() and indexConcurrent().

In the app/Http/Controllers/ directory, create a DashboardController.php file with the following content:

```
<?php

namespace App\Http\Controllers;

class DashboardController extends Controller
{
    public function index()
    {
```

```
        return view('welcome');
    }
    public function indexConcurrent()
    {
        return view('welcome');
    }
}
```

We just created the controller's methods, so they are just loading the view. Now we are going to add some logic in the methods, creating a query in the model file and calling it from the controllers.

Creating the query

To allow the controller to load data, we are going to implement the **query**, the logic that retrieves data from the events table. To do that, we are going to use the query scope mechanism provided by Laravel. The query scope allows you to define the logic in the model and reuse it in your application.

The query scope we are going to implement will be placed in the scopeOfType() method in the Event model class. The scopeOfType() method allows you to extend the functionalities of the Event model and add a new method, ofType():

```php
<?php

namespace App\Models;

use Illuminate\Database\Eloquent\Factories\HasFactory;
use Illuminate\Database\Eloquent\Model;

class Event extends Model
{
    use HasFactory;

    /**
     * This is a simulation of a
     * complex query that is time-consuming
     *
     * @param  mixed  $query
     * @param  string  $type
     * @return mixed
```

```
    */
    public function scopeOfType($query, $type)
    {
        sleep(1);
        return $query->where('type', $type)
        ->where('description', 'LIKE', '%something%')
        ->orderBy('date')->limit(5);
    }
}
```

The Event model file is located in the app/Models directory. The file is Event.php.

The query returns the event type defined as an argument ($type) and selects the rows where the description contains the word something (through the 'LIKE' operator).

In the end, we are going to sort the data by date (orderBy) and limit it to five records (limit).

In order to highlight the benefits of the optimizations we are going to implement, I am going to add a 1-second sleep function to simulate a time-consuming operation.

The DashboardController file

Now we can open again the DashboardController file and implement the logic to call the four queries – the first one for counting the events:

```
Event::count();
```

The second one is for retrieving the events with the defined query via the ofType function for events with the 'INFO' type:

```
Event::ofType('INFO')->get();
```

The third one is for retrieving the 'WARNING' event:

```
Event::ofType('WARNING')->get();
```

The last one is for retrieving the 'ALERT' event:

```
Event::ofType('ALERT')->get();
```

Let's put it all together in the controller `index()` method to call the queries sequentially:

```
use App\Models\Event;
// …
public function index()
{
    $time = hrtime(true);
    $count = Event::count();
    $eventsInfo = Event::ofType('INFO')->get();
    $eventsWarning = Event::ofType('WARNING')->get();
    $eventsAlert = Event::ofType('ALERT')->get();
    $time = (hrtime(true) - $time) / 1_000_000;

    return view('dashboard.index',
        compact('count', 'eventsInfo', 'eventsWarning',
                'eventsAlert', 'time')
    );
}
```

The `hrtime()` method is used for measuring the execution time of all four queries.

Then, after all the queries are executed, the `dashboard.index` view is called.

Now, in the same way, we will create the `indexConcurrent()` method, where the queries are executed in parallel via the `Octane::concurrently()` method.

The `Octane::concurrently()` method has two parameters. The first one is the array of tasks. A task is an anonymous function. The anonymous function can return a value. The `concurrently()` method returns an array of values (the returned values of the task array). The second parameter is the amount of time in milliseconds that `concurrently()` waits for the completion of the task. If a task takes more time than the second parameter (milliseconds), the `concurrently()` function will raise a `TaskTimeoutException` exception.

The implementation of the `indexConcurrent()` method is located in the `DashboardController` class:

```
public function indexConcurrent()
{
    $time = hrtime(true);
    try {
        [$count, $eventsInfo, $eventsWarning, $eventsAlert] =
```

```
        Octane::concurrently([
            fn () => Event::count(),
            fn () => Event::ofType('INFO')->get(),
            fn () => Event::ofType('WARNING')->get(),
            fn () => Event::ofType('ALERT')->get(),
        ]);
    } catch (TaskTimeoutException $e) {
        return "Error: " . $e->getMessage();
    }
    $time = (hrtime(true) - $time) / 1_000_000;


    return view('dashboard.index',
        compact('count', 'eventsInfo', 'eventsWarning',
                'eventsAlert', 'time')
    );
}
```

To use TaskTimeoutException correctly, you have to import the class:

```
use Laravel\Octane\Exceptions\TaskTimeoutException;
```

The last thing you have to implement to render the pages is the view.

Rendering the view

In the controller, the last instruction of each method is returning the view:

```
return view('dashboard.index',
        compact('count', 'eventsInfo', 'eventsWarning',
                'eventsAlert', 'time')
        );
```

The view() function loads the resources/views/dashboard/index.blade.php file (dashboard.index). To share data from the controller to the view, we are going to send some arguments to the view() function, such as $count, $eventsInfo, $eventsWarning, $eventsAlert, and $time.

The view is an HTML template that uses Blade syntax to show variables such as $count, $eventsInfo, $eventsWarning, $eventsAlert, and $time:

```
<x-layout>
    <div>
        Count : {{ $count }}
    </div>
    <div>
        Time : {{ $time }} milliseconds
    </div>
    @foreach ($eventsInfo as $e)
    <div>
        {{ $e->type }} ({{ $e->date }}): {{ $e->description }}
    </div>
    @endforeach


    @foreach ($eventsWarning as $e)
    <div>
        {{ $e->type }} ({{ $e->date }}): {{ $e->description }}
    </div>
    @endforeach


    @foreach ($eventsAlert as $e)
    <div>
        {{ $e->type }} ({{ $e->date }}): {{ $e->description }}
    </div>
    @endforeach
</x-layout>
```

The view inherits the layout (via the x-layout directive) so you can create the resources/views/components/layout.blade.php file:

```
<html>
    <head>
        <title>{{ $title ?? 'Laravel Octane Example' }}
        </title>
        <meta charset="UTF-8">
```

```
    <meta name="viewport" content="width=device-width,
        initial-scale=1.0">

    </head>
    <body>
        <h1>Laravel Octane Example</h1>
        <hr/>
        {{ $slot }}
    </body>
</html>
```

Now you have the data in your database, the query in the `model` class, and the controller that loads the data via the model and sends data to the view, and the view template file.

We also have two routes: the first one is `/dashboard` with sequential queries, and the second one is `/dashboard-concurrent` with parallel queries.

Just for this example, the query is forced to take 1 second (in the model method).

If you open your browser at `http://127.0.0.1:8000/dashboard`, you will see that each request takes more than 3 seconds (each query takes 1 second). This is the sum of all the execution times of each query.

If you open your browser at `http://127.0.0.1:8000/dashboard-concurrent`, you will see that each request takes 1 second to be executed. This is the maximum execution time of the most expensive query.

This means that you have to call multiple queries in your controller to retrieve data. To render the page, you can use the `Octane::concurrently()` method.

The `Octane::concurrently()` method is also great in other scenarios (not just loading data from a database), such as making concurrent HTTP requests. So, in the next section, we are going to use the `Octane::concurrently()` method to retrieve data from HTTP calls (instead of retrieving data from a database). Let's see how.

Making parallel HTTP requests

Think about the scenario in which you have to add a new web page in your application, and to render the web page, you have to call more than one API because you need multiple pieces of data from multiple sources (list of products, list of news, list of links, etc.) for the same page. In the scenario with one web page that needs data from multiple API calls, you could perform the HTTP requests simultaneously to reduce the response time of the page.

For this example, to simplify the explanation, we will avoid using the controller and the view. We are going to collect JSON responses from APIs and then we will merge the responses into one JSON response. The important aspect to focus on is the mechanism of calling HTTP requests to third-party HTTP services because our goal is to understand how to make the HTTPS call concurrently.

To simulate the HTTP service, we are going to create two new routes:

- `api/sentence`: An API endpoint that replies with a JSON with a random sentence
- `api/name`: An API endpoint that replies with a JSON with a random first name

Both endpoint APIs implement a `sleep()` function of 1 second to allow the client (who calls the endpoint) to wait for the answer. This is a way to simulate a slow API and see the benefit we can obtain from parallel HTTP requests.

In the `routes/web.php` file, you can add the two routes that implement the APIs:

```
Octane::route('GET', '/api/sentence', function () {
    sleep(1);
    return response()->json([
        'text' => fake()->sentence()
    ]);
});
Octane::route('GET', '/api/name', function () {
    sleep(1);
    return response()->json([
        'name' => fake()->name()
    ]);
});
```

Now, using the `Http::get()` method to perform HTTP requests, you can implement the logic to retrieve data from two APIs sequentially:

```
Octane::route('GET', '/httpcall/sequence', function () {
    $time = hrtime(true);
    $sentenceJson =
      Http::get('http://127.0.0.1:8000/api/sentence')->
      json();
    $nameJson =
      Http::get('http://127.0.0.1:8000/api/name')->json();
    $time = hrtime(true) - $time;
```

```
    return response()->json(
        array_merge(
            $sentenceJson,
            $nameJson,
            ["time_ms" => $time / 1_000_000]
        )
        );
});
```

Using `Octane::concurrently()`, you can now call the two `Http::get()` methods, using the HTTP request as `Closure` (anonymous function), as we did for the database queries:

```
Octane::route('GET', '/httpcall/parallel', function () {
    $time = hrtime(true);
    [$sentenceJson, $nameJson] = Octane::concurrently([
        fn() =>
            Http::get('http://127.0.0.1:8000/api/sentence')->
            json(),
        fn() =>
            Http::get('http://127.0.0.1:8000/api/sequence')->
            json()
    ]
    );
    $time = hrtime(true) - $time;
    return response()->json(
        array_merge(
            $sentenceJson,
            $nameJson,
            ["time_ms" => $time / 1_000_000]
        )
        );
});
```

If you open your browser to `http://127.0.0.1:8000/httpcall/sequence`, you will see that the response time is more than 2,000 milliseconds (the sum of the execution time of the two sleep functions, and some milliseconds for executing the HTTP connection).

If you open your browser to http://127.0.0.1:8000/httpcall/parallel, you will see that the response takes more than 1,000 milliseconds (the two HTTP requests are performed in parallel).

Using Octane::concurrently() could help you save some total response time when making these examples with database queries or fetching external resources.

Managing HTTP errors

While executing HTTP calls in parallel, you have to expect that, sometimes, the external service could answer with an error (for example, with an HTTP status code 500). For better error management in the source code, we must also properly deal with the case where we get an empty response from the API, which typically results in a response with errors (for example the API returns a status code 500).

Here, we demonstrate that we are going to implement an API that returns 500 as an HTTP status code (an internal server error message):

```
Octane::route('GET', '/api/error', function () {
    return response(
        status: 500
    );
});
```

Then, we can call the API error route in one of our concurrent HTTP calls. If we are not managing the error, we will receive an error such as this one:

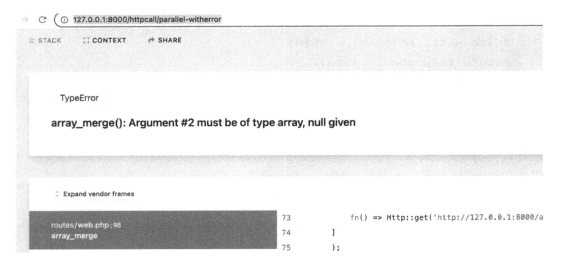

Figure 4.7: The unmanaged error in the browser

So, we could improve our code by managing the following:

- The exception that could come from the execution of concurrent HTTP calls
- The empty response value with the Null coalescing operator
- Initializing the arrays as an empty array

In the routes/web.php file, we can improve the API calls and make them more reliable:

```
Route::get('/httpcall/parallel-witherror', function () {
    $time = hrtime(true);
    $sentenceJson = [];
    $nameJson = [];
    try {
        [$sentenceJson, $nameJson] = Octane::concurrently([
            fn () => Http::get(
                'http://127.0.0.1:8000/api/sentence')->json()
                ?? [],
            fn () => Http::get(
                'http://127.0.0.1:8000/api/error')->json() ??
                [],
        ]
        );
    } catch (Exception $e) {
        // The error: $e->getMessage();
    }
    $time = hrtime(true) - $time;

    return response()->json(
        array_merge(
            $sentenceJson,
            $nameJson,
            ['time_ms' => $time / 1_000_000]
        )
    );
});
```

In this way, if an exception is raised or we receive an HTTP error as a response, our software will manage these scenarios.

The suggestion is that even if you are focusing on performance aspects, you don't have to lose focus on the behavior of the application and managing the unhappy paths correctly.

Now that we understand how to execute tasks in parallel, we can focus on caching the response to avoid calling external resources (database or web service) for every request.

Understanding the caching mechanism

Laravel provides the developer with a strong mechanism for caching.

The caching mechanism can be used with a provider chosen from the database, Memcached, Redis, or DynamoDB.

Laravel's caching mechanism allows data to be stored for later retrieval quickly and efficiently.

This is very useful in cases where retrieving data from an external service with a database or web service can be a time-consuming operation. After information retrieval, storing the retrieved information in a cache mechanism is possible to make future information retrieval easier and faster.

So basically, a caching mechanism exposes two basic functionalities: caching of information and retrieval from the cache of information.

To properly retrieve information each time a cached item is used, it is appropriate to use a storage key. This way, it is possible to cache a lot of information identified by a specific key.

Laravel's caching mechanism, through the special `remember()` function, allows retrieving a piece of information tied to a specific key. If this information has become obsolete because the storage time-to-live has been exceeded, or if the key is not cached, then the `remember()` method allows calling an anonymous function that has the task of getting the data from the external resource, which can be the database or a web service. Once the original data is retrieved, the `remember()` function automatically returns the data but, at the same time, also takes care of caching it with the user-defined key.

Here is an example of using the `remember()` function:

```
use Illuminate\Support\Facades\Cache;


$secondsTimeToLive = 5;
$cacheKey= 'cache-key';
$value = Cache::remember($cacheKey, $secondsTimeToLive,
function () {
    return Http::get('http://127.0.0.1:8000/api/sentence')
```

```
        ->json() ?? [];
});
```

The `remember()` functionality applied to each HTTP request in the previous example can be implemented in an anonymous function:

```
$getHttpCached = function ($url) {
        $data = Cache::store('octane')->remember(
                'key-'.$url, 20, function () use ($url) {
            return Http::get(
               'http://127.0.0.1:8000/api/'.$url)->json() ??
               [];
        });

        return $data;
    };
```

The anonymous function can then be invoked by the `Octane::concurrently()` function for each concurrent task:

```
[$sentenceJson, $nameJson] = Octane::concurrently([
            fn () => $getHttpCached('sentence'),
            fn () => $getHttpCached('name'),
        ]
        );
```

So, the final code in a route in the `routes/web.php` file is as follows:

```
Octane::route('GET','/httpcall/parallel-caching', function () {
    $getHttpCached = function ($url) {
        $data = Cache::store('octane')->remember(
                'key-'.$url, 20, function () use ($url) {
            return Http::get(
               'http://127.0.0.1:8000/api/'.$url)->json() ??
               [];
        });

        return $data;
    };
```

```
    $time = hrtime(true);
    $sentenceJson = [];
    $nameJson = [];
    try {
        [$sentenceJson, $nameJson] = Octane::concurrently([
            fn () => $getHttpCached('sentence'),
            fn () => $getHttpCached('name'),
        ]
        );
    } catch (Exception $e) {
        // The error: $e->getMessage();
    }
    $time = hrtime(true) - $time;

    return response()->json(
        array_merge(
            $sentenceJson,
            $nameJson,
            ['time_ms' => $time / 1_000_000]
        )
    );
});
```

The following are some considerations about the code:

- We used the Octane route (faster than Laravel routes).

- The $url parameter of the anonymous function is used to create the cache key and to call the right API via Http::get().

- We used the cache with Octane as the driver, Cache::store('octane').

- We used the remember() function for the cache.

- We set the time-to-live of the cache item at 20 seconds. It means that after 20 seconds, the cache item is generated, and the code provided by the anonymous function will be called.

This code improves the response time dramatically thanks to the cache.

However, the code could be more optimized.

We cache the result from each HTTP request. But, we could cache the result provided by Octane::concurrently. So, instead of caching each HTTP request, we could cache the result that comes from Octane::concurrently(). This allows us to save more time by avoiding the execution of Octane::concurrently() if the value is cached.

In this case, we can move Octane::concurrently() in the body of the anonymous function called by remember():

```
Octane::route('GET', '/httpcall/caching', function () {
    $time = hrtime(true);
    $sentenceJson = [];
    $nameJson = [];
    try {
        [$sentenceJson, $nameJson] =
        Cache::store('octane')->remember('key-checking',
                                        20, function () {
            return Octane::concurrently([
                fn () => Http::get(
                  'http://127.0.0.1:8000/api/sentence')->
                  json(),
                fn () => Http::get(
                  'http://127.0.0.1:8000/api/name')->json(),
            ]);
        });
    } catch (Exception $e) {
        // The error: $e->getMessage();
    }
    $time = hrtime(true) - $time;

    return response()->json(
        array_merge(
            $sentenceJson,
            $nameJson,
            ['time_ms' => $time / 1_000_000]
        )
    );
});
```

In this case, from the log of the requests, you can see that the APIs are only called the first time, then the data is retrieved from the cache, and the execution time is reduced:

```
200      GET /api/sentence ........ 18.57 mb 17.36 ms
200      GET /api/name ........... 18.57 mb 17.36 ms
200      GET /httpcall/caching .... 17.43 mb 59.82 ms
200      GET /httpcall/caching ..... 17.64 mb 3.38 ms
200      GET /httpcall/caching ..... 17.64 mb 2.36 ms
200      GET /httpcall/caching ..... 17.64 mb 3.80 ms
200      GET /httpcall/caching ..... 17.64 mb 3.30 ms
```

The first call to the caching route takes around 60 milliseconds; the subsequent requests are much faster (around 3 milliseconds)

If you try to do the same test by calling the HTTP requests sequentially and not using the cache, you will see higher values as response times. You will also see that the API will be called every time, making the speed and the reliability of your application dependent on a third-party system because the reliability and the speed depend on the way the third-party system (that provides the APIs) creates the response.

For example, by calling HTTP requests sequentially, with no cache – even if the APIs are provided by Octane (so in a faster way) – you will obtain the following:

```
200      GET /api/sentence ........ 18.57 mb 15.22 ms
200      GET /api/name ............ 18.68 mb 0.64 ms
200      GET /httpcall/sequence ... 18.79 mb 60.81 ms
200      GET /api/sentence ........ 18.69 mb 3.26 ms
200      GET /api/name ............ 18.69 mb 1.68 ms
200      GET /httpcall/sequence ... 18.94 mb 15.55 ms
200      GET /api/sentence ........ 18.70 mb 1.30 ms
200      GET /api/name ............ 18.70 mb 1.09 ms
200      GET /httpcall/sequence .... 18.97 mb 9.52 ms
200      GET /api/sentence ........ 18.71 mb 1.32 ms
200      GET /api/name ............ 18.71 mb 1.05 ms
200      GET /httpcall/sequence .... 19.00 mb 9.28 ms
```

While you might think that this is not a great improvement or that these values are machine-dependent, a small improvement (our response time has gone from 10-15 milliseconds to 2-3 milliseconds) for a single request could have a big impact, especially if, in a production environment, you have a huge number of simultaneous requests. The benefit of each small improvement for a single request is multiplied by the number of requests you might have in a production environment with many concurrent users.

Now that we understand a bit more about caching, we could refactor our dashboard by adding the cache for event retrieval.

Refactoring the dashboard

We are going to create a new route, /dashboard-concurrent-cached, with the Octane route and we are going to call a new DashboardController method, indexConcurrentCached():

```php
// Importing Octane class
use Laravel\Octane\Facades\Octane;
// Importing Response class
use Illuminate\Http\Response;
// Importing the DashboardController class
use App\Http\Controllers\DashboardController;

Octane::route('GET', '/dashboard-concurrent-cached', function
() {
    return new Response((new DashboardController)->
      indexConcurrentCached());
});
```

In the controller app/Http/Controllers/DashboardController.php file, you can add the new method:

```php
public function indexConcurrentCached()
{
    $time = hrtime(true);
    try {
        [$count,$eventsInfo,$eventsWarning,$eventsAlert] =
        Cache::store('octane')->remember(
            key: 'key-event-cache',
            ttl: 20,
            callback: function () {
                return Octane::concurrently([
                    fn () => Event::count(),
                    fn () => Event::ofType('INFO')->get(),
                    fn () => Event::ofType('WARNING')->
                            get(),
                    fn () => Event::ofType('ALERT')->get(),
```

```
                ]);
            }
        );
    } catch (Exception $e) {
        return 'Error: '.$e->getMessage();
    }
    $time = (hrtime(true) - $time) / 1_000_000;

    return view('dashboard.index',
        compact('count', 'eventsInfo', 'eventsWarning',
                'eventsAlert', 'time')
    );
}
```

In the new method, we do the following:

- Call the `remember()` method to store the values in the cache

- Execute `Octane:concurrently` to parallelize the queries

- Use `'key-event-cache'` as the key name for the cache item

- Use 20 seconds as the cache time-to-live (after 20 seconds, the queries will be executed and retrieve new values from the database)

- Use the same query of the `/dashboard` route and the same blade view (to make a good comparison)

Now, you can restart your Octane worker with `php artisan octane:reload` if you are not using the automatic reloader (as explained in *Chapter 2, Configuring the RoadRunner Application Server*), and then access the following:

- `http://127.0.0.1:8000/dashboard` to load the page with sequential queries and without a caching mechanism

- `http://127.0.0.1:8000/dashboard-concurrent-cached` to load the page with parallel queries and with a caching mechanism

Now that we have implemented the logic and opened the pages, we are going to analyze the result.

The result

The result that you can see is impressive as, from a response that took more than 200 milliseconds, you will now have a response that takes 3 or 4 milliseconds.

The longer response is from /dashboard, where sequential queries are implemented without a cache. The fastest responses come from /dashboard-concurrent-cached, which uses Octane::concurrently() to execute the queries, and the result is cached for 20 seconds:

```
200     GET /dashboard ...... 19.15 mb 261.34 ms
200     GET /dashboard ...... 19.36 mb 218.45 ms
200     GET /dashboard ...... 19.36 mb 223.23 ms
200     GET /dashboard ...... 19.36 mb 222.72 ms
200     GET /dashboard-concurrent-cached
....................... 19.80 mb 112.64 ms
200     GET /dashboard-concurrent-cached .....................
........... 19.81 mb 3.93 ms
200     GET /dashboard-concurrent-cached .....................
........... 19.81 mb 3.69 ms
200     GET /dashboard-concurrent-cached ....................
........... 19.81 mb 4.28 ms
200     GET /dashboard-concurrent-cached .....................
........... 19.81 mb 4.62 ms
```

When you are caching data in Octane Cache, you should also be aware of the cache configuration. A wrong configuration could raise some errors in your application.

The cache configuration

A typical exception that you might see when you start to use Octane Cache in a real scenario is something like this:

```
Value [a:4:{i:0;i:100000;i:...] is too large for [value] column
```

The solution to the error message above is to change the cache configuration by increasing the number of bytes allocated for storing the cache values. In the config/octane.php file, you can configure the cache for the number of rows and the number of bytes allocated for the cache.

By default, the configuration is as follows:

```
'cache' => [
    'rows' => 1000,
    'bytes' => 10000,
],
```

If you get the `Value is too large` exception in your browser, you might have to increase the number of bytes in the `config/octane.php` file:

```
'cache' => [
    'rows' => 1000,
    'bytes' => 100000,
],
```

So now, using Octane features, you can improve the response time and some aspects of your application.

Summary

In this chapter, we built a very simple application that allowed us to cover multiple aspects of building a Laravel application, such as importing the initial data, optimizing the routing mechanism, integrating third-party data via HTTP requests, and using a cache mechanism via Octane Cache. We also used some Laravel Octane features in order to reduce the page loading response time thanks to the following:

- `Octane::route` for optimizing the routing resolution process
- `Octane::concurrently` for optimizing and starting parallel tasks
- Octane Cache for adding a cache based on Swoole to our application

We learned how to execute queries and API calls concurrently and use the cache mechanism for reusing the content across the requests.

In the next chapter, we will take a look at some other aspects of performance that are not strictly provided by Octane but can affect your Octane optimization process.

We will also apply a different strategy for caching using the scheduled tasks provided by Octane and other optimizations.

5
Reducing Latency and Managing Data with an Asynchronous Approach

In the previous chapter, we created a new Laravel Octane application, and we applied some features provided by Laravel Octane to improve performance and reduce the response time of our application.

In this chapter, we will try to optimize more things, such as access to a database and changing and improving the caching strategy. To improve and make queries to data faster, we will explain the benefit that comes from indexing columns. For caching, we will also take a look at a cache-only approach.

For example, in the previous chapter, we executed queries in parallel. Now, we will optimize a query because parallelizing something fast is better than parallelizing something slower. Query optimization allows code to retrieve data from databases faster, reducing the latency that normally occurs when reading data from any source (a file, database, or network).

Then, we will show how to make the query process faster using a caching mechanism. We will use a *cache-only* approach, which means that the code will always retrieve data from a cache. There is a task that performs queries and stores the result in the cache. Therefore, the pre-caching mechanism is totally independent of the code that needs data. For this reason, we will refer to this approach as asynchronous because the running code that needs data has not had to wait for the process that retrieves data and then fill the cache.

The goal of this chapter is to reduce HTTP request response times. To achieve a reduction in response issues, we will see how to implement information retrieval through query optimization and a caching mechanism that separates cache retrieval times from cache fill times.

In this chapter, we will cover the following topics:

- Optimizing the queries with indexes
- Making the cache mechanism asynchronous

Technical requirements

We will assume that you have the application set up from the previous chapter.

You need to set up your Laravel Octane application with the event migration and event seeder required for this chapter. The requirement for the current chapter is to have PHP 8 installed or, if you want to use a container approach, you have to install Docker Desktop (`https://www.docker.com/products/docker-desktop/`) or a similar tool to run the Docker images.

> **Source code**
>
> You can find the source code of the examples used in this chapter in the official GitHub repository of this book: `https://github.com/PacktPublishing/High-Performance-with-Laravel-Octane/tree/main/octane-ch05`.

Optimizing queries with indexes

In the previous chapter, we made queries in parallel.

What if the parallelized queries were slow? Most of the time, implementing an optimization means acting on multiple aspects. In the previous chapter, we saw how to parallelize queries. This approach, as we saw, brings great benefits, but there is something more we can do. What we want to achieve is to further reduce the latency of each individual parallelized task when retrieving data.

To do just that, what we are going to do now is optimize each query that we parallelized in the previous chapter and explore the reasoning behind each one.

We are going to analyze what the characteristics of the query are and what fields are involved in the rows selection phase and the sorting phase.

Let's start with the following example query:

```
return $query->where('type', $type)
    ->where('description', 'LIKE', '%something%')
    ->orderBy('date')->limit(5);
```

We can see that in the query, we are performing some operations on some columns (filtering by type, filtering by description, etc.).

To make the query faster, on the columns involved in the query, we are going to create indexes. Indexes in a database are a data structure used by the database engine when a query is executed.

To use an analogy to explain indexes, it's as if you want to look up a word in a dictionary. Starting on the first page and scrolling down each successive page, you get to the term you are looking for. The time it takes to find the term depends on the number of words and the placement of the word. Just think what it is like to find a word that begins with the letter z in a vocabulary of thousands and thousands of words.

An index in a database is like having an index in a dictionary, where each letter has a page number. Using an index, the access to a term is much more immediate. Nowadays, various databases have a very complex and performant index system, so the aforementioned analogy is straightforward compared to reality. Still, it allows us to understand how much the existence of an index on a field used for searching or sorting is crucial for performance.

In Laravel, if you want to create an index, you can do it in a migration file. A migration file is a file where you can define the structure of your database table. In the migration file, you can list the columns of your table and define the type of the columns (string, integer, date, time, etc.).

In *Chapter 4*, *Building a Laravel Octane Application*, we already created the structure of the `events` table (the table used for our examples). The goal now is to analyze which columns could benefit from index creation, and we will see how to create indexes in the migration file.

In the migration file created in the previous chapter (in the `database/migrations/` directory), we created a table with some fields:

```
Schema::create('events', function (Blueprint $table) {
    $table->id();
    $table->foreignIdFor(User::class)->index();
    $table->string('type', 30);
    $table->string('description', 250);
    $table->integer('value');
    $table->dateTime('date');
    $table->timestamps();
});
```

Some of these fields were used for filtering the rows.

For example, in the model file in `app/Models`, we implemented the following query:

```
return $query->where('type', $type)
    ->where('description', 'LIKE', '%something%')
    ->orderBy('date')->limit(5);
```

This means the fields used in the query for filtering are `'type'` and `'description'`. The field used for sorting is the `'date'` field.

Therefore, we are going to create three indexes (one for the `'type'` column, one for the `'description'` column, and one for the `'date'` column) in one migration.

Before creating the indexes, let's look at the response time of the dashboard controller, just to have a baseline value so that we can later check the improvement in terms of time saved. The dashboard controller is the place where the query is called via the `ofType()` method:

```
$count = Event::count();
$eventsInfo = Event::ofType('INFO')->get();
$eventsWarning = Event::ofType('WARNING')->get();
$eventsAlert = Event::ofType('ALERT')->get();
```

To show the response time of the dashboard controller with all of these queries, you can start Laravel Octane via the following command:

php artisan octane:start

Then, you can access it with your web browser at `http://127.0.0.1:8000/dashboard` and see the response time in the console.

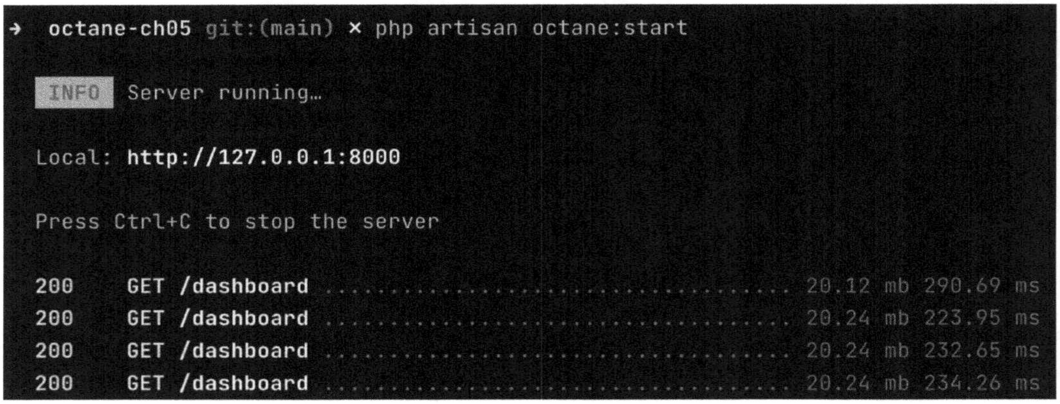

Figure 5.1: The response time of the dashboard controller without the index usage

As you can see, the response time is more than 200 milliseconds.

Now, we are going to create indexes, and we will see the new response time.

Creating indexes

We can create a new migration with the make:migration command:

php artisan make:migration create_event_indexes

Then, in the yyyy_mm_dd_hhMMss _create_event_indexes.php file, created in the database/migrations/ directory, with the up() method, we are going to use the index() method to create an index for each column:

```
Schema::table('events', function (Blueprint $table) {
    $table->index('type', 'event_type_index');
    $table->index('description',
                  'event_description_index');
    $table->index('date', 'event_date_index');
});
```

The first parameter of the index() method is the column name; the second one is the index name. The index name is useful when, for example, you want to drop the column in the down() method. The down() method is used in case of rollback:

```
Schema::table('events', function (Blueprint $table) {
    $table->dropIndex('event_type_index');
    $table->dropIndex('event_description_index');
    $table->dropIndex('event_date_index');
});
```

To apply the newly created indexes, you have to run the migration via the migrate command:

php artisan migrate

If you want to check whether everything is fine, you can use the db:table command and see whether the new indexes are listed:

php artisan db:table events

If the indexes are created, you will see them listed in the **Index** section:

```
→  octane-ch05 php artisan db:table events

   events ..................................................................................................
   Columns ............................................................................................. 8
   Size ........................................................................................... 0.02MiB

   Column ............................................................................................ Type
   id autoincrement, bigint, unsigned ............................................................. bigint
   user_id bigint, unsigned ....................................................................... bigint
   type string .................................................................................... string
   description string ............................................................................. string
   value integer ................................................................................. integer
   date datetime ................................................................................ datetime
   created_at datetime, nullable ................................................................ datetime
   updated_at datetime, nullable ................................................................ datetime

   Index ..................................................................................................
   event_date_index date ................................................................................
   event_description_index description ..................................................................
   event_type_index type ................................................................................
   events_user_id_index user_id .........................................................................
   PRIMARY id ........................................................................... unique, primary
```

Figure 5.2: Executing db:table can show the new indexes

Now that the indexes are created, we are going to analyze the query already implemented that uses the following fields: `type`, `description`, and `date` for filtering and sorting. The query that we are going to use is the query implemented in the `scopeOfType()` method in the `Event` model (in the `app/Models/Event.php` file):

```
public function scopeOfType($query, $type)
{
    return $query->where('type', $type)
    ->where('description', 'LIKE', 'something%')
    ->orderBy('date')->limit(5);
}
```

To see the different response times with the index usage, after you have created the indexes, with your web browser again, you can access the dashboard controller via `http://127.0.0.1:8000/dashboard` and see the results in the console:

```
→  octane-ch05 git:(main) × php artisan octane:start

   INFO  Server running…

   Local: http://127.0.0.1:8000

   Press Ctrl+C to stop the server

   200    GET /dashboard ................................ 20.12 mb 101.03 ms
   200    GET /dashboard ................................ 20.24 mb 35.37 ms
   200    GET /dashboard ................................ 20.24 mb 46.59 ms
   200    GET /dashboard ................................ 20.24 mb 46.99 ms
   200    GET /dashboard ................................ 20.24 mb 33.85 ms
   200    GET /dashboard ................................ 20.24 mb 48.00 ms
```

Figure 5.3: The response time of the dashboard controller with database indexes

If you want to obtain more analytical metrics from the benefit of index usage, you can use some tools provided directly by the database. For example, in the case of MySQL, you can access MySQL Command Prompt (with php artisan db, as explained in the next few lines), execute the query, and retrieve the Last_query_cost value. You will obtain a value representing the query execution cost, based on the number of operations performed.

In order to compare the last query costs, we are going to execute a query with indexes first and then without. In the example, we are going to extract the Last_query_cost metric.

This is how we do it:

1. Make sure that you are using the latest version of your migration:

    php artisan migrate

2. Then, open the MySQL command line with the db command:

    php artisan db

    The db command executes the MySQL client according to Laravel configuration (database name, username, password, and table name).

3. In MySQL Command Prompt, you can execute the query on the events table:

    SELECT * FROM events WHERE description LIKE 'something%';

4. Once you complete the query, the result will be shown.

5. Then, execute the following:

    SHOW STATUS LIKE 'Last_query_cost';

6. You will now see the metrics that represent the query costs (dependent on the number of operations performed by the query on the data):

```
mysql> SHOW STATUS LIKE 'Last_query_cost';
+-----------------+-----------+
| Variable_name   | Value     |
+-----------------+-----------+
| Last_query_cost | 5.209000  |
+-----------------+-----------+
1 row in set (0.00 sec)
```

7. If you now try to drop the indexes, a rollback on the migration will occur (because the latest step of our migration is index creation via the rollback command):

```
php artisan migrate:rollback --step=1
```

8. Then, in the MySQL command prompt, execute the following again:

```
SELECT * FROM events WHERE description LIKE 'something%';
```

9. And then, execute this:

```
SHOW STATUS LIKE 'Last_query_cost';
```

You will see the following:

```
mysql> SHOW STATUS LIKE 'Last_query_cost';
+-----------------+-------------+
| Variable_name   | Value       |
+-----------------+-------------+
| Last_query_cost | 1018.949000 |
+-----------------+-------------+
1 row in set (0.01 sec)
```

As you can see, without indexes, you have higher query costs.

Without indexes, things get worse if you perform more complex queries like the query we have in your Event model, which is filtering using the type and description fields and sorting by date.

Let's try to execute the following query:

```
SELECT * FROM events WHERE type='ALERT' AND description LIKE
'something%' ORDER BY date;
```

And then, after you have executed the query, you ask MySQL to show the `Last_query_cost` metric without indexes:

```
mysql> SHOW STATUS LIKE 'Last_query_cost';
+-----------------+---------------+
| Variable_name   | Value         |
+-----------------+---------------+
| Last_query_cost | 10645.949000  |
+-----------------+---------------+
```

Then, you ask MySQL to show the `Last_query_cost` metric with indexes:

```
mysql> SHOW STATUS LIKE 'Last_query_cost';
+-----------------+------------+
| Variable_name   | Value      |
+-----------------+------------+
| Last_query_cost | 16.209000  |
+-----------------+------------+
```

As you can see, the differences are enormous.

We have now compared the queries with and without indexes. With the acquired knowledge, we can improve the response time of the query by fine-tuning the types of indexes we are using.

In the query, we are filtering the rows with the `description` column that starts with a specific word. In the query, we are filtering all the descriptions that start with the word `something`. But what if we want to filter all the rows with the `description` column that includes the word `something`? In our model in our previous chapter, in order to select all the descriptions that include a specific word, we used the `LIKE` operator with a wildcard:

```
description LIKE '%something%'
```

However, if we want to optimize the query in order to reduce its response time, specifically for text, we have another powerful database feature for filtering and searching text – **full-text indexes**.

Creating full-text indexes

What we are going to do is to change the standard index for the `description` column to a full-text index, and we will see how it performs. This is how we do it:

1. Create a new migration:

    php artisan make:migration create_event_fulltext_index

    In the `yyyy_mm_dd_hhMMss_create_event_fulltext_index.php` file, in the `database/migrations/` directory, in the `up()` method, we are going to drop the previous index and create a new full text.

2. In the `up()` method, we have to drop the previous index on the `description` column and then create a full-text index via the `fullText()` method:

```
Schema::table('events', function (Blueprint $table) {
    $table->dropIndex('event_description_index');
    $table->fullText('description', 'event_description_
fulltext_index');
});
```

Regarding which method is faster, the `LIKE` operator used to filter the description column that starts with a specific term (`'something%'`) is faster than a full-text search, but it only covers filtering a column that starts with a specific word. Full-text execution is quicker than a `LIKE` approach that searches with wildcards and is more powerful, especially if you want to search for one more term. Let's try to use the `whereFullText()` method in the query:

```
public function scopeOfType($query, $type)
{
    return $query->where('type', $type)
        //->where('description', 'LIKE', '%something%')
        ->whereFullText('description', 'something other')
        ->orderBy('date')->limit(5);
}
```

The `whereFullText()` method is used with two parameters; the first one is the `description` parameter, which is the column name to filter, and the second parameter is the string to search for.

Now that we have a full-text search in place, we can add another improvement – caching the result of the queries.

Optimized queries and caching

At this point, we have seen how to optimize queries through parallel execution, the use of a cache, and by applying indexes on search fields.

However, if we optimize queries by caching the result, the first query, or queries for which the cached result is obsolete or deleted, will have to deal with the cost of loading data from the database to update and refresh the cache.

With the caching strategy we are about to discuss, we will try to prevent cache refresh times from affecting the response time of our application.

This scenario can be optimized by changing the caching strategy to having queries in a request using only the values coming from a cache (a cache-only approach), and creating a process that takes care of the retrieval of the results and their caching. This process operates asynchronously and is decoupled from the queries generated by the requests.

This approach allows for improved performance because it tends to eliminate slow queries, and that should take care of data retrieval because the periodic refresh of the cache is done through an external command. We can add the external command that is executed every *n* second and retrieves new data, and then fill the cache. All requests get data from the cache. To execute the interval command, we can use another Octane functionality, the `tick()` method. Let's see how.

Making the cache mechanism asynchronous

In *Chapter 3, Configuring the Swoole Application Server*, we explored the `Octane::tick()` method.

The `tick()` method allows you to execute a function every *n* seconds.

The caching strategy could be reviewed by delegating data loading to a specific function. This specific function is responsible for retrieving data with the query from the database (and not from the cache), and once the data is retrieved, the function stores the results in the cache. The function is called via the `Octane::tick()` method and executed – for example, maybe every 60 seconds, fresh data from the database is retrieved, and it fills the cache. All the requests retrieve the data from the cache.

With the asynchronous caching strategy, all the requests retrieve data from the cache.

The cache is refreshed by the task called via `tick()`.

To implement the asynchronous caching strategy, we are doing the following:

1. Implementing the `tick()` function in the application service provider
2. Storing the result in the Octane Cache
3. Implementing the controller that reads the cache
4. Implementing the routing

Implementing the tick() function in the application service provider

In order to launch the caching task when the framework is bootstrapped, we can set the `tick()` method in the App Service Provider. The App Service Provider is a file called while the framework is instanced. So, in the `app/Providers/AppServiceProvider.php` file, in the `boot()` method, you have to implement the `Octane::tick()` function:

```
// including these classes
use Illuminate\Support\Facades\Log;
use Laravel\Octane\Facades\Octane;
use App\Models\Event;
use Illuminate\Support\Facades\Cache;

// in the boot() method
Octane::tick('caching-query', function () {
    Log::info('caching-query.', ['timestamp' => now()]);
    $time = hrtime(true);
    $count = Event::count();
    $eventsInfo = Event::ofType('INFO')->get();
    $eventsWarning = Event::ofType('WARNING')->get();
    $eventsAlert = Event::ofType('ALERT')->get();
    $time = (hrtime(true) - $time) / 1_000_000;
    $result = ['count' => $count,
        'eventsInfo'=> $eventsInfo,
        'eventsWarning' => $eventsWarning,
        'eventsAlert'=> $eventsAlert,
    ];

    Cache::store('octane')->put('cached-result-tick', $result);
})
->seconds(60)
->immediate();
```

In the `tick()` function, we are going to execute all queries and then store the result in the Octane Cache (specifying the `octane` store): `Cache::store('octane')->put()`.

The two essential methods are seconds(), where we can define the cadence, the interval in seconds, and immediate(), which sets an immediate execution of the tick() function.

Every 60 seconds, the queries are executed automatically, and the result is stored in the cache.

Implementing the controller that reads the cache

Now that we have implemented the tick() event that fills the cache, we can focus on the controller, where we can load data from the cache.

The method for retrieving data from the cache is Cache::store('octane')->get(). With the Cache::store('octane') method, you retrieve the Cache instance provided by Octane. With the get() method, you will retrieve the value stored in the cache. Here's the code that retrieves the value from the cache in the app/Http/Controllers/DashboardController.php file:

```
use Illuminate\Support\Facades\Cache;
use Exception;

public function indexTickCached()
{
    $time = hrtime(true);
    try {
        $result = Cache::store('octane')->get(
            'cached-result-tick');
    } catch (Exception $e) {
        return 'Error: '.$e->getMessage();
    }
    $time = (hrtime(true) - $time) / 1_000_000;
    $result['time'] = $time;

    return view('dashboard.index', $result);
}
```

In the controller, as you can see, there is an asynchronous approach because there are no more pending operations that depend on the databases, and we are delegating loading from the database to an external function. The only dependency in the controller is with the cache.

Implementing routing

In the `routes/web.php` file, you can add a new route:

```
use Laravel\Octane\Facades\Octane;
use App\Http\Controllers\DashboardController;
use Illuminate\Http\Response;


Octane::route('GET', '/dashboard-tick-cached', function () {
    return new Response((new DashboardController)->
                          indexTickCached());
});
```

As already shown in previous chapters, we can optimize the routing loading with `Octane::route()`, which helps to reduce the response time. As you can see, we are using `Octane::route()` for performance reasons, and then we set the path as `/dashboard-tick-cached` and call the `indexTickCached()` method.

Showing the results

If we open the browser to the initial dashboard route, where the queries were not optimized and not cached, and then open the browser to the new route, where the queries are cached, we can see a massive difference in terms of response time:

```
200     GET /dashboard ........... 19.71 mb 66.60 ms
200     GET /dashboard ........... 19.83 mb 42.31 ms
200     GET /dashboard ........... 19.83 mb 37.31 ms
200     GET /dashboard ........... 19.83 mb 30.07 ms
200     GET /dashboard ........... 19.83 mb 42.25 ms
200     GET /dashboard-tick-cached . 19.89 mb 7.51 ms
200     GET /dashboard-tick-cached . 19.89 mb 4.32 ms
200     GET /dashboard-tick-cached . 19.89 mb 4.96 ms
```

As you can see, the dashboard path has a response time of 30–40 milliseconds. The `dashboard-tick-cached` route is around 5 milliseconds.

It is a significant improvement, and again, when you think about performance, you have to think in terms of the impact of this improvement on thousands of requests.

This brings us to the end of the chapter.

Summary

In this chapter, we have seen how combining the caching mechanism, routing optimization, an asynchronous approach, and query optimization can benefit information retrieval. The caching mechanism combined with the asynchronous approach helps us reduce the data retrieval response time for every request (even if the cache is outdated). The query optimization reduces the time spent retrieving fresh data to fill a cache. The routing optimization helps us to save more milliseconds when the frameworks resolve the routes, reducing response time.

In the next chapter, we will try to address situations where we need to perform a time-consuming task – that is, operations that take some time to complete but where we cannot use caching mechanisms.

Caching mechanisms can be beneficial when retrieving information. On the other hand, if we need to perform a task such as writing, sending, or transforming data, we most likely will need to use some other tool. In the next chapter, we will see how.

Part 4: Speeding Up

This part shows how to configure tools to support Laravel Octane in a performant architecture. Tools such as queues and how to set up the system for the production environment are explained through practical examples. This part comprises the following chapters:

- *Chapter 6, Using Queues to Apply the Asynchronous Approach in Your Application*
- *Chapter 7, Configuring the Laravel Octane Application for the Production Environment*

6

Using Queues to Apply the Asynchronous Approach in Your Application

In the previous chapter, we saw how delegating some tasks to external functions in the controller while creating the HTTP request can have a positive impact from a performance perspective. However, the case analyzed in the previous chapter was limited to a scoped topic: querying the database and populating the cache. This strategy is also known as the *cache-only strategy*. The process only needs the data to be retrieved from the cache.

This type of approach works in a case where information needs to be retrieved from a data source. Typically, applications are more complex than this, such as when executing specific tasks that need to modify and process data.

Think about a scenario where there is a request that starts a backup process. Typically, a backup process takes time to be completed. Implementing a synchronous approach means that the controller (who serves the request) keeps the client (the web page) on hold until the process is completed. Two bad things about this solution are that the user sees a long waiting loader in the browser, and the solution could fall into the *request time-out* scenario.

For asynchronous implementation, an additional tool is typically used to manage the list of task execution requests. To allow asynchronous implementation, we need a mechanism that works as a queue where we have a producer that needs the job to be executed (the producer produces jobs requests and feeds the queue) and a consumer that extracts the job requests from the queue and performs the job, one at a time. Typically, a queue system adds some features to monitor the queues, manage the queues (empty the queues), manage prioritization, and manage multiple channels or queues.

Laravel provides a mechanism to implement all the queue logic management, such as putting a task in the queue, extracting a job from a queue, managing failed tasks, and notifying the user about executions.

To store the list of tasks, Laravel allows the developer to select one of the queue backends available: the database, Redis, Amazon SQS, or beanstalkd.

This chapter aims to help you understand what a queuing mechanism is, how it can be used with Laravel, and how it is configured – because an asynchronous approach with queues not only reduces the response time but also implements a different user experience, especially when time-consuming tasks have to be handled.

In this chapter, we will cover the following:

- Introducing the queue mechanism in Laravel
- Installing and configuring the queue
- Managing the queues
- Managing queues with Redis and monitoring them

Technical requirements

Thanks to the previous chapters, we assume you have your basic Laravel application installed with Laravel Octane. For the current chapter, you can use Octane with Swoole, Open Swoole, or RoadRunner.

The source code for the example described in the current chapter is available here: `https://github.com/PacktPublishing/High-Performance-with-Laravel-Octane/tree/main/octane-ch06`.

Introducing the queue mechanism in Laravel

We will implement a simple use case in order to shed light on the asynchronous aspects and how much a queuing mechanism can improve the user experience for the end users of our web application.

For every request, a time-consuming task will be executed. To simulate the time-consuming task, we will call `sleep()` to last 3 seconds. The `sleep()` function, which suspends execution for a certain number of seconds, is intended to simulate the execution of a task that may take some time to implement. In a real case, the `sleep()` function is replaced with complex business logic that could take a certain amount of time to complete.

With the synchronous approach, the request will hold the response to the browser for 3 seconds. As a user, you will request the page, wait for 3 seconds, and then the page will be shown. The page will contain the message that the operation is completed – so you are safe and sure that the process is correctly executed in 3 seconds, but you have to wait for the answer.

With the asynchronous approach, the routing mechanism takes charge of the request; a job is created in the queue to call the logic, which includes a call to the `sleep(3)` function to simulate a time-consuming operation, provided by the `ProcessSomething::handle()` function. After the job is created in the queue, the response is generated and sent to the client.

The user will receive a response in a few milliseconds without waiting for the task to be completed. You know that the task has been pushed into the queue, and some workers will execute the job.

To carry out the asynchronous approach, we are going to do the following:

1. Install the queue mechanism, creating a table in the database to store the job queued.

2. Create the class for implementing the logic of the job.

3. Implement a time-consuming logic in the `handle()` method of the job class.

4. Create a route for calling the `handle()` method in the classical synchronous way.

5. Create another route for asynchronously calling the job through the queue mechanism.

6. Analyze the result of calling the two routes.

First, we will create the data structure that will allow the queue mechanism to store the list of jobs. We can use a MySQL database or Redis. To simplify understanding, we will initially use the MySQL database-based queuing mechanism (because it is more basic and simpler). Subsequently, we will use the more advanced Redis-based system.

Redis

Redis is an open source data store used for storing data, caching values, and queuing data/messages. Working mainly in memory, one of its main features is speed.

Now, we are going to configure the database-based queuing mechanism in a Laravel application.

Installing and configuring the queue

To create the data structure to store the jobs in the queue, we can execute the command in the terminal in our Laravel project directory:

```
php artisan queue:table
```

The command is provided by Laravel without the need to install additional packages.

The `queue:table` command creates a new migration file for creating the `jobs` table.

The file is created in the `database/migrations/` directory.

```
return new class extends Migration
{
    /**
     * Run the migrations.
     *
     * @return void
     */
    public function up()
    {
        Schema::create('jobs', function (Blueprint $table) {
            $table->bigIncrements('id');
            $table->string('queue')->index();
            $table->longText('payload');
            $table->unsignedTinyInteger('attempts');
            $table->unsignedInteger('reserved_at')->nullable();
            $table->unsignedInteger('available_at');
            $table->unsignedInteger('created_at');
        });
    }
```

Figure 6.1: The migration file for creating the jobs table

The migration will create the following:

- A new table named 'jobs'
- 'id': For the unique identifier
- 'queue': The queue name, helpful for controlling the queue via the command line
- 'payload': The data in JSON format that contains information to manage and launch the task by the consumer of the queue
- 'attempts': The number of attempts to execute the jobs
- 'reserved_at': The timestamp when the task is taken in charge by the consumer
- 'available_at': When the task is available to be consumed
- 'created_at': When the job is created in the queue

An example of the JSON payload (in the payload field) is as follows:

```
{
    "displayName" : "App\\Jobs\\ProcessSomething",
    "failOnTimeout" : false,
    "retryUntil" : null,
    "data" : {
        "command" :
            "O:25:\"App\\Jobs\\ProcessSomething\":0:{}",
        "commandName" : "App\\Jobs\\ProcessSomething"
    },
    "maxExceptions" : null,
    "maxTries" : null,
    "uuid" : "e8b0c6c7-29ce-4108-a74c-08c70bb679a6",
    "timeout" : null,
    "backoff" : null,
    "job" : "Illuminate\\Queue\\CallQueuedHandler@call"
}
```

The JSON payload has some attributes:

- failOnTimeout: The Boolean field indicating whether the job should fail when it times out
- retryUntil: The timestamp (integer field) indicating when the job should time out
- maxExceptions: The number (integer field) of times to attempt a job after an exception
- maxTries: The number (integer field) of times to attempt a job
- uuid: The UUID (string field) of the job
- timeout: The number (integer field) of seconds the job can run
- backoff: The number of seconds to wait before retrying a job that encountered an uncaught exception – can be an array of integers to track the seconds for each attempt (a job could be attempted more than once because of errors)
- job: The name (string field) of the queued job class

Once we have created the schema definition for the table with the php artisan queue:table command, we can create the jobs table in the database via the migrate command.

For creating the table in the database, you can launch the `migrate` command:

```
php artisan migrate
```

To check that the correct table has been created, you can use the `db:table` command:

```
php artisan db:table jobs
```

Now that all the data structure is in place, we have to create the files to implement the logic of our jobs.

Managing the queues

For managing the jobs, in Laravel, by convention, we have one class for each job. The job class has to implement the `handle()` method. The method `handle()` is invoked by the framework when the job has to be executed.

For creating the class to manage the jobs, see the following:

```
php artisan make:job ProcessSomething
```

With the `make:job` command, a new file, `app/Jobs/ProcessSomething.php`, that includes the `ProcessSomething` class with some methods ready to be filled with the logic is created. The primary methods are the constructor and the method invoked for managing the job, the `handle()` method.

We will implement the logic into the `handle()` method. In the `app/Jobs/ProcessSomething.php` file, insert the following:

```
public function handle()
{
    Log::info('Job processed START');
    sleep(3);
    Log::info('Job processed    END');
}
```

As an example of a time-consuming operation, we are going to pause the execution of the `handle()` method for 3 seconds via the `sleep()` function. This means that the thread will be suspended for 3 seconds. We will log the method execution's start and end to track more information. You can find the logs in the `storage/logs/laravel.log` file with the default configuration.

The classical synchronous approach

Typically, with a synchronous approach, if you call the method, the response takes 3 seconds.

In the `routes/web.php` file, we are going to add a new `/time` route for dispatching (requesting the execution of) the job synchronously:

```
Route::get('/time-consuming-request-sync', function () {
    $start = hrtime(true);
    ProcessSomething::dispatchSync();
    $time = hrtime(true) - $start;

    return view('result', [
        'title' => url()->current(),
        'description' => 'the task has been complete with
                          dispatchSync()',
        'time' => $time,
    ]);
});
```

Calling the static `dispatchSync()` method allows you to invoke the `handle()` method (through the queue mechanism) synchronously. This is the classic scenario we have in PHP when we call a method.

To render the view, we have to implement the *result* view, a basic blade template to render the title, description, and time set in the `view()` function.

Create the `resources/views/result.blade.php` blade file:

```
<!DOCTYPE html>
<html lang="{{ str_replace('_', '-', app()->getLocale()) }}">
<head>
  <meta charset="utf-8">
  <meta name="viewport" content="width=device-width,
    initial-scale=1">
  <title>Laravel</title>
</head>
<body class="antialiased">
  <div class="relative flex items-top justify-center
    min-h-screen bg-gray-100 dark:bg-gray-900
    sm:items-center py-4 sm:pt-0">
```

```html
<div class="max-w-6xl mx-auto sm:px-6 lg:px-8">
    <div class="mt-8 bg-white dark:bg-gray-800
        overflow-hidden shadow sm:rounded-lg">
        <div class="grid grid-cols-1">
            <div class="p-6">
                <div class="flex items-center">
                    <div class="ml-4 text-lg leading-7
                        font-semibold">{{ $title}}</div>
                </div>
                <div class="ml-12">
                    <div class="mt-2 text-gray-900
                        dark:text-gray-900 text-2xl">
                        {{ $description }}
                    </div>
                    <div class="mt-2 text-gray-900
                        dark:text-gray-900 text-2xl">
                        {{ $time / 1_000_000 }} milliseconds
                    </div>
                </div>
            </div>
        </div>
    </div>
</div>
</body>
</html>
```

The blade file will show the result on the page displaying the $description and $time parameters. If you request the page via your web browser at http://localhost:8000/time-consuming-request-sync, you must wait at least 3 seconds before the page is fully rendered. This figure shows that the time value is slightly more than 3,000 milliseconds:

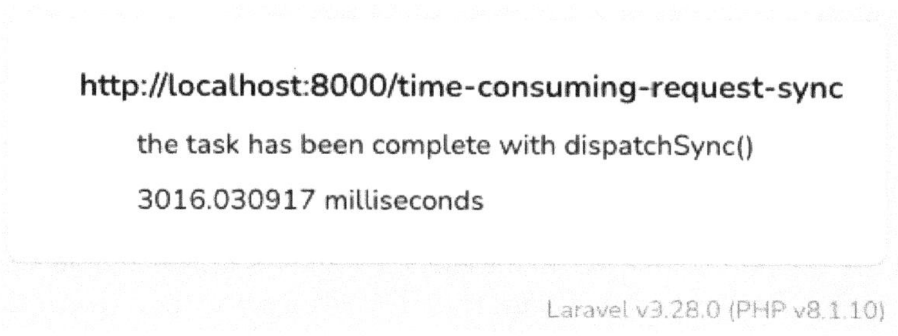

http://localhost:8000/time-consuming-request-sync

the task has been complete with dispatchSync()

3016.030917 milliseconds

Laravel v3.28.0 (PHP v8.1.10)

Figure 6.2: The synchronous job execution

This means your job takes 3 seconds, and Laravel waits to send the response until the job is completed.

This is probably nothing special because, as PHP developers, we are used to managing synchronous jobs. Even if we are not managing queues, the PHP engine synchronously drives the methods and functions. Some other languages have the async call for functions – so now, let's see how to dispatch the execution to another process that can asynchronously execute the job.

The asynchronous approach

Instead of using the sync method, we will dispatch the jobs through the queue.

To extract the jobs from the queue, it is necessary to start a specific process for the consumer to take over the various tasks. To begin the process, you can execute the artisan command:

```
php artisan queue:work
```

The command checks whether any jobs are in the jobs table and *consumes* the queue by deleting the row after the job is completed.

From the *producer* side, in the router logic, we can call the dispatch() functions available:

```
Route::get('/time-consuming-request-async', function () {
    $start = hrtime(true);
    dispatch(new ProcessSomething());
// OR you can use ProcessSomething::dispatch();
    $time = hrtime(true) - $start;

    return view('result', [
        'title' => url()->current(),
        'description' => 'the task has been queued',
```

```
        'time' => $time,
    ]);
});
```

The `dispatch()` functions need the instance of the job class as a parameter. The `dispatch()` function will send the job instance to the queue and release the control to the caller, with no need to wait for the complete execution of the job. In the end, the response is created immediately without waiting for the job to be completed, instead of the typical behavior where the response is created once the job is completed.

We are just sending an instance of the `ProcessSomething` class to the `dispatch()` method. The convention is that the *consumer* will execute the `handle()` method when taking care of the job.

Now, you can open the browser and call the URL, `http://localhost:8000/time-consuming-request-async`:

http://127.0.0.1:8000/time-consuming-request-async

the task has been queued

6.805292 milliseconds

Laravel v9.28.0 (PHP v8.1.10)

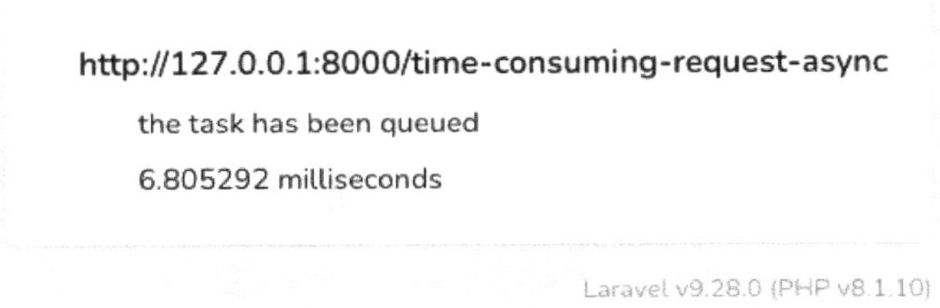

Figure 6.3: The asynchronous execution

To allow the browser to receive the response, we have to be sure that we launched two commands:

- `php artisan octane:start`: For launching the Octane server, listening on port `8000`
- `php artisan queue:work`: For launching the *consumer* service, for executing the jobs

If you want to see the status of the queue, you can execute it via the command line:

```
php artisan queue:monitor default
```

The `queue:monitor` command will show the status of the queues and the jobs in the queue for each queue:

```
→ octane-ch06 git:(main) × php artisan queue:monitor default

Queue name ........................................ Size / Status
default ...................................................... [0] OK
```

Figure 6.4: The queue monitor tool

If you have pending jobs in the queue, you will see the number of waiting jobs in the square brackets:

```
→  octane-ch06 git:(main) × php artisan queue:monitor default

   Queue name ..................................... Size / Status
   default .............................................. [76] OK
```

Figure 6.5: The queue with some waiting jobs

In the example, we have 76 jobs in the queue.

If you use a database as a backend for managing queues, I suggest increasing your confidence directly by querying the jobs table with SQL.

You can access the database with the `artisan db` command:

```
php artisan db
```

Then, you can execute SQL queries; for example, you could count how many rows each queue has:

```
select count(id) as jobs_count, queue
from jobs
group by queue;
```

The SQL command counts the number of identifiers (the count) from the `jobs` table, grouping the rows by the `queue` field.

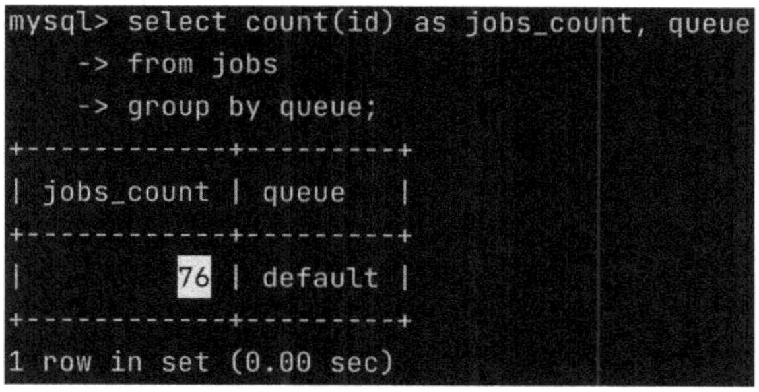

```
mysql> select count(id) as jobs_count, queue
    -> from jobs
    -> group by queue;
+------------+---------+
| jobs_count | queue   |
+------------+---------+
|         76 | default |
+------------+---------+
1 row in set (0.00 sec)
```

Figure 6.6: Executing queries on the jobs table

If you have more than one job in the queue, please be sure that you are running the consumer:

```
php artisan queue:work
```

If you want to have more than one consumer that executes your jobs in parallel, you can launch `queue:work` more than once. If you launch `queue:work` twice, you will have two consumers that extract the jobs from the queue.

If you have to manage time-consuming tasks, using a queue is not just implementing an asynchronous approach to manage tasks. It is a way to control the level of parallelism and contain the number of concurrent time-consuming tasks that your architecture can take charge of or handle. By managing time-consuming tasks synchronously, if there were a high number of requests, you could reach a high number of concurrent requests on the web server, and your system could collapse due to the high resource usage.

Delegating tasks to specific workers means that you keep the load on the workers used for serving the requests lighter, and you can launch the consumer processes on a dedicated instance or virtual machine. You can also increase the number of processes for the consumers.

Managing multiple queues

You can use more than one queue – for example, `first` and `second`.

When you have to assign the job to a queue, you can use the `onQueue()` method:

```
ProcessSomething::dispatch()->onQueue("first");
```

You can control both queues:

```
php artisan queue:monitor first,second
```

Then, you can launch the consumer for the `"first"` queue:

```
php artisan queue:work --queue=first
```

And then for the queue named `"second"`:

```
php artisan queue:work --queue=second
```

You can, for example, launch two consumers for the `"first"` queue and only one for the `"second"` queue, giving more priority to the `"first"` queue (because it has two dedicated consumers instead of one). To achieve this, in different shell environments, you can launch the following:

```
php artisan queue:work --queue=first
php artisan queue:work --queue=first
php artisan queue:work --queue=second
```

If you need to clear the queue and delete all pending tasks, you can use the `queue:clear` command:

```
php artisan queue:clear database --queue=first
```

Here, `database` is the name of the connection (we are using the database now; in the next section, we will use another type of connection), and we can also define the queue via the `--queue` parameter. You can also specify more than one queue:

```
php artisan queue:clear database --queue=first,second
```

The connection is optional; if we don't specify `database` (the connection) on the command line, the `QUEUE_CONNECTION` environment parameter (from `.env` file) will be used.

Now we have seen how to create and manage queues with the database as the backend, let's try to configure Redis as the backend.

Managing queues with Redis and monitoring them

Using `database` as the connection is convenient for people who start using the queue and have already had the database set up for storing application data, for example. Why use Redis instead of a database? Because Redis has more optimization for managing queues than a database, and you can use Laravel Horizon to monitor the queue. Laravel Horizon provides a web dashboard for monitoring your queues and the metrics about the usage of the queues.

As the first step of managing queues with Redis, first, let's install the Redis service.

Installing Redis

Installing Redis means that you have added software and service to your stack. If you are a macOS user, you can install it via **Homebrew**:

```
brew install redis
brew services start redis
```

The first command installs the software; the second one starts the service.

You can use your package manager if you have a GNU/Linux distribution; Redis is included in most GNU/Linux distributions.

Alternatively, you can use Sail (as shown previously in *Chapter 3, Configuring the Swoole Application Server*, in the *Setting up Laravel Sail* section, during the installation of Swoole).

Execute the following `sail:install` command:

```
php artisan sail:install
```

Be sure you also select the Redis service (option number **3**).

```
Which services would you like to install? [mysql]:
  [0] mysql
  [1] pgsql
  [2] mariadb
  [3] redis
  [4] memcached
  [5] meilisearch
  [6] minio
  [7] mailhog
  [8] selenium
> 3
```

Figure 6.7: The Laravel Sail configuration for Redis

In the case that you need more than one service (for example, both MySQL and Redis), you can select 0, 3 (comma-separated).

Once the Redis service runs, we can start configuring the queue mechanism to use Redis as the connection.

Configuring Redis

In the environment configuration file placed in the Laravel project directory (the `.env` file), we have to adjust some parameters:

```
QUEUE_CONNECTION=redis
REDIS_CLIENT=predis
REDIS_HOST=localhost
REDIS_PASSWORD=null
REDIS_PORT=6379
```

With `QUEUE_CONNECTION`, you define the connection to be used (`redis`).

With `REDIS_CLIENT`, you can define the client to be used by Laravel to connect to the Redis service. The default is `phpredis` (which uses the PECL module), or you can use `predis`, which uses the PHP package: `https://github.com/predis/predis`. The `phpredis` module is written in C, so could be faster than a `predis` implementation (which is pure PHP). On the other hand, a `predis` implementation has a lot of features and the community and the development team are really supportive.

If you want to change the name of the default queue (typically, "default"), you can add the REDIS_QUEUE parameter into the .env file:

```
REDIS_QUEUE=yourdefaultqueue
```

About the connection between Laravel and Redis service, if you are not familiar with Redis, my suggestion is to start by using the predis package just because it requires adding a package:

composer require predis/predis

If you are installing the predis/predis package, you have to set the REDIS_CLIENT parameter to predis:

```
REDIS_CLIENT=predis
```

All of these configurations are related to the .env file.

If you are using the Predis package to complete the configuration correctly, you must set an alias in the Laravel bootstrap configuration. To do that, in the config/app.php file, in the 'aliases' section, add a specific entry for Redis:

```
'aliases' => Facade::defaultAliases()->merge([
    // 'ExampleClass' => App\Example\ExampleClass::class,
    'Redis' => Illuminate\Support\Facades\Redis::class,
])->toArray(),
```

Setting the alias is useful for Laravel to access the Redis object correctly and helps Laravel to correctly resolve the reference to the Redis object because we could have a name conflict with the Redis object provided by the Predis package and with the Redis object provided by the phpredis module. If you forget to set this configuration, you will not receive an error in executing the Laravel application, but you could face some unexpected application behavior. For example, when you want to clear the queue on a specific connection, the queue selected is not removed.

If you switch the connection, for example, from the database to Redis, you don't have to change anything in your application code. The Laravel queue mechanism provides an abstraction layer that hides all the different implementations specific to each connection (under the hood, managing a database connection uses other implementations than a Redis connection).

If you are using the code (in the routes/web.php file), see the following:

```
Route::get('/time-consuming-request-async', function () {
    $start = hrtime(true);
    ProcessSomething::dispatch()->onQueue("first");
    $time = hrtime(true) - $start;
```

```
    return view('result', [
        'title' => url()->current(),
        'description' => 'the task has been queued',
        'time' => $time,
    ]);
});
```

We are using the "first" queue on the configured connection.

To see the configured connection, you can use the about command:

php artisan about --only=drivers

The command displays the information on the drivers used by your Laravel application for the queue:

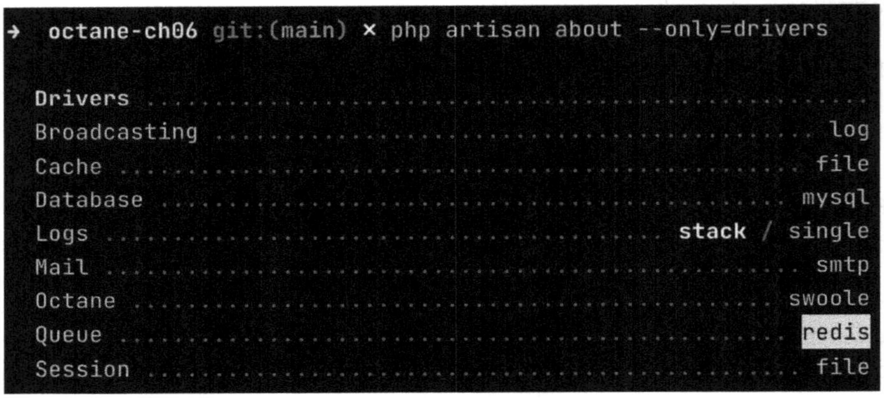

Figure 6.8: The driver configuration

The command shows that for the queue, we are using Redis as the backend connection for queues.

Now, opening the page at http://127.0.0.1:8000/time-consuming-request-async, you will see that deferring the task to a Redis queue is faster than a database. For example, the dispatch method takes less than a millisecond on my local machine. With the database connection in our example, the same code takes 7 milliseconds:

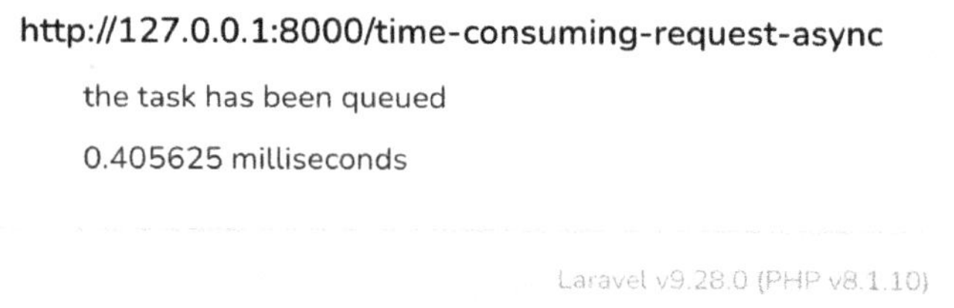

http://127.0.0.1:8000/time-consuming-request-async

the task has been queued

0.405625 milliseconds

Laravel v9.28.0 (PHP v8.1.10)

Figure 6.9: Dispatching a job on the Redis queue

With the queue mechanism, we improved the *responsiveness* of the application, allowing the user to receive an immediate response about the task. Then, with the Redis connection, we reduced the time taken to queue the job.

Thanks to the Redis connection, we can also use Laravel Horizon for monitoring the queues.

Monitoring queues with Laravel Horizon

We are going to add Horizon to our application. This means that you can access the Horizon dashboard using the /horizon path at the end.

To install Laravel Horizon, you have to install the package:

```
composer require laravel/horizon
```

Then ,you need to publish the files needed by Horizon: the configuration file (config/horizon.php), the assets (in the public/vendor/horizon/ directory), and the service provider (in the app/Providers/HorizonServiceProvider.php file):

```
php artisan horizon:install
```

horizon:install will copy all needed files into the correct directories.

Launch Laravel Octane with the following:

```
php artisan octane:start
```

With the default Horizon configuration, you can access `http://127.0.0.1:8000/horizon`:

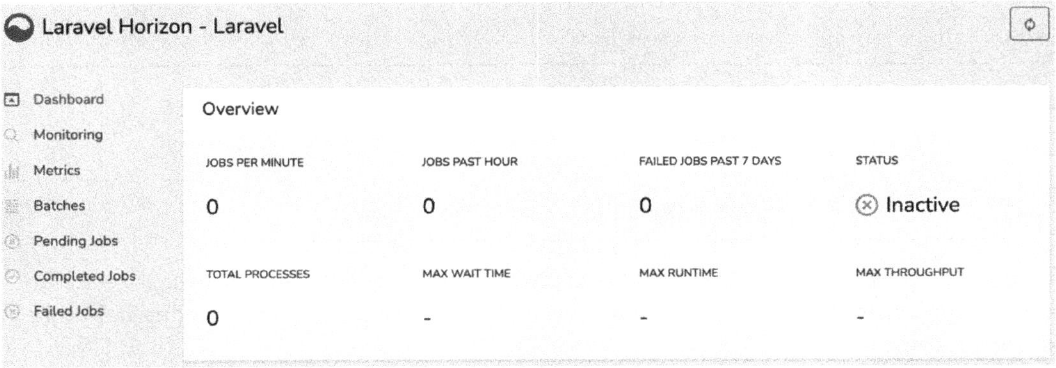

Figure 6.10: The Laravel Horizon dashboard

The status indicates whether the supervisor of Horizon processes is up and running. To collect all the metrics correctly, launch the supervisor:

```
php artisan horizon
```

Then, if you load the page `http://127.0.0.1:8000/time-consuming-request-async` multiple times, multiple jobs are created in the queue. Launch the consumer:

```
php artisan queue:work --queue=first
```

You will have a consumer up and running, ready to execute jobs from the queue.

So, you have Octane, Horizon, and the queue worker up and running. Load the page multiple times, and then go to the dashboard at `http://127.0.0.1:8000/horizon/dashboard`. You will see the dashboard page filled with the metrics:

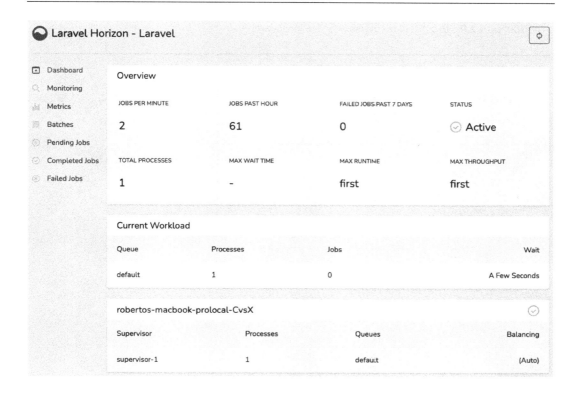

Figure 6.11: The Horizon Dashboard page elaborating on the metrics

The metrics are retrieved and calculated in **near real-time** (**NRT**), which means that under the hood, there is a process that continuously (every 3 seconds) performs HTTP calls to the /horizon/api/ stats endpoint implemented by Horizon.

If you look at the API's response, you could programmatically retrieve the same information you can see in the dashboard UI.

> **Note**
>
> In a scenario such as this one, where you have the UI that calls multiple endpoints (APIs) via polling (every 3 seconds), if you have the APIs served by Octane, you can inherit all the benefits that come from using Octane. Octane reduces latency thanks to all its optimization.

For the statistics API, the JSON response has this structure:

```json
{
    "failedJobs": 0,
    "jobsPerMinute": 1,
    "pausedMasters": 0,
    "periods": {
        "failedJobs": 10080,
        "recentJobs": 60
    },
    "processes": 1,
    "queueWithMaxRuntime": "first",
    "queueWithMaxThroughput": "first",
    "recentJobs": 0,
    "status": "running",
    "wait": {
        "redis:default": 0
    }
}
```

Here, you can retrieve the following:

- The count of recently failed jobs, `"failedJobs"`.
- The jobs processed per minute since the last snapshot, `"jobsPerMinute"`.
- The number of master supervisors that are currently paused, `"pausedMasters"`.
- The configuration for how long (in minutes) Horizon has managed the recent and failed jobs. The values are expressed in seconds, and the configuration is defined in `config/horizon.php` in the `trim` section.
- The process count, `"processes"`.
- The name of the queue that has the most extended runtime, `"queueWithMaxRuntime"`.
- The name of the queue that has the highest throughput, `"queueWithMaxThroughput"`.
- The count of recent jobs, `"recentJobs"`.
- The status of the supervisor (the supervisor is the process run via `php artisan horizon`), `"status"`.
- The time to clear per queue, `"wait"` (in seconds).

The Horizon dashboard is a convenient way to monitor the status and statistics of all running queues. Via Horizon, you can't control the queues; for managing the queues, you can use the commands explained here:

- `queue:monitor`: For monitoring the status of the queue and the number of waiting jobs
- `queue:clear`: For deleting all the jobs in a connection or queue
- `queue:flush`: For deleting all failed jobs
- `queue:forget`: For deleting a specific failed job (to avoid retrying the execution of a failed job)
- `queue:retry`: Retrying the execution of a previously failed job
- `queue:work`: Processing the queue (or the queues specified via the -queue parameter)

Thanks to all these commands, you can control the queue's status and health.

With Horizon, you can monitor the queues' execution and status.

So now, we have an architecture for running and monitoring queues with Redis as the backend.

Summary

With an asynchronous approach, we can defer the execution of some tasks and be more responsive to the user's request. In other words, the job is queued, and it will be taken care of later. The behavior is different from the classic approach where the job is executed immediately, but the benefit is that control of the application's UI is consistently available to the user. The user experience of your application is smoother, so the user, in the meantime, can do other things with your application. Besides improving the user experience, the asynchronous approach is more scalable because you can control the processes that will take charge of the jobs in a granular way. You also can execute more than one worker via the `php artisan queue:work` command – and if your hardware architecture has more virtual machines for running backend processes, you can run the consumer processes across multiple virtual machines.

To achieve asynchronous architecture in the current chapter, we introduced the queue mechanism; we have shown the following:

- How to install and set up the queue in Laravel first with a database as the backend connection, then with a more powerful backend with Redis
- The differences between executing a job in a synchronous way and in an asynchronous way
- The benefit of using the queues in terms of the responsiveness of the system and the impact on the user experience
- Installing Horizon to monitor the queue usage

In the next chapter, we will see how to prepare the application and how to set up the tools to deploy a Laravel Octane application in a production environment.

7

Configuring the Laravel Octane Application for the Production Environment

In previous chapters and this book in general, we have seen how to use Laravel Octane and implement several optimizations to improve the performance and responsiveness of our application. We operated mainly in the local development environment. In addition, we have worked at the application level.

This chapter will address system configurations, setup, and fine-tuning in a production environment. Understanding the specific configurations for the production environment is very useful, especially because the configuration of a local environment is typically different from the configuration of the production environment due to the different settings and services available.

In practice, we will cover the following:

- A typical production architecture
- How the reference architecture can be optimized
- Which specific Laravel operations can bring benefits from a performance perspective
- The deployment strategy via **Makefile**

Technical requirements

We start with the assumption that the reader has a production environment available. Typically, this kind of environment can be on a virtual machine, such as one (or more) Amazon EC2 instances, or an environment provided by cloud PaaS solutions, such as *Platfrom.sh*, or by cloud servers/service providers, such as *Digital Ocean* or *Vultr*.

We will not address the specific installation of these environments; however, we will give indications of a typical architecture that combines the use of a web server such as nginx on one (or more) GNU/Linux-type virtual machines.

The source code and the configuration files of the examples described in the current chapter are available here: `https://github.com/PacktPublishing/High-Performance-with-Laravel-Octane/tree/main/octane-ch07`.

Introducing the production architecture

In the local environment development scenario (as in the previous chapters), it is fine to use the `php artisan octane:start` command to start Laravel Octane. However, it may be helpful to configure a different architecture in a production environment where the requirements are different from a local environment.

Typically, in a local development environment, you need automation and configurations that are useful for developing new features, such as automatic service reload. In contrast, in a production environment, the configuration must allow for optimum levels of performance and reliability. Therefore, choosing a different, production-specific architecture in this production scenario is common.

The design of the production environment architecture we are going to address involves the use of a web server to initially sort HTTP requests. The web server's task will be to differentiate requests for static assets (where no PHP engine involvement is required) from requests that require server-side processing. Usually, a request for a page, which requires server-side processing, might be followed by other requests for static assets such as CSS files, JavaScript files, and image files. PHP involvement is not needed for these types of files, so the web server's task will be to serve these types of assets immediately, in the fastest way possible. On the other hand, for requests requiring PHP execution, the web server will forward the request to the service provided by Laravel Octane.

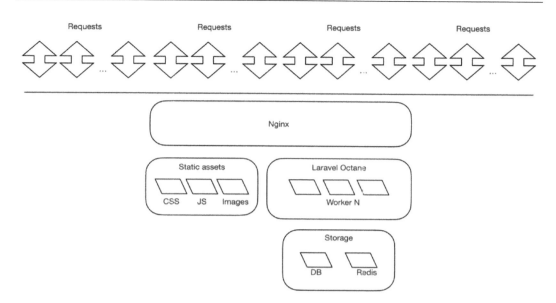

Figure 7.1: A typical production architecture

To configure the production environment, several steps must be performed:

1. Place the static assets in a specific directory (we can organize the asset types into multiple subfolders).

2. Configure the nginx (`nginx.conf` file) for acting as a proxy.

3. Configure nginx to forward the page requests to Laravel Octane.

4. Set nginx to listen to the HTTPS protocol.

5. Set the communication between nginx and Laravel Octane as standard HTTP.

6. All the URL generated by Laravel has to be set up as served as HTTPS (for absolute URL generation).

Let's start with managing public assets.

Managing public static assets

If you look at the directory structure of a Laravel application, you will notice that there is a public directory, the purpose of which is to contain static files.

The public directory also contains the built CSS and JS files (by the `npm run build` command). If your application needs a CSS style, you will probably set it in the `resources/css/app.css` file (or in the `resources/js` directory for the JavaScript files). Once you run `npm run build`, the CSS and JS files are built, and the output files are stored in the `public/build` directory.

To provide static files in HTTP responses, it is not necessary to involve the PHP engine. Avoiding involving the PHP engine to provide static files minimizes the use of resources such as CPU and memory and, consequently, also reduces response times.

Static files typically are files with a specific extension such as jpg, jpeg, gif, css, png, js, ico, or html. For static files, we could need a particular configuration – for example, for turning off the log, defining an expiration date, or adding a header. If you need to set a specific setting for the assets, the suggestion is to set a particular configuration for nginx in the server section in the nginx.conf file (or the file for your domain):

```
location ~* \.(jpg|jpeg|gif|css|png|js|ico|html)$ {
    access_log off;
    expires max;
    log_not_found off;
    add_header X-Debug-Config STATICASSET;
}
```

For files with assets extensions, I added a header (header in the HTTP response) named X-Debug-Config, for debugging purposes. The configuration was set to check whether STATICASSET value was present in the HTTP response headers.

Setting nginx as a proxy

In the production environment, we are going to use nginx to reply to HTTP requests. We are going to configure nginx as a proxy because, for the PHP files, nginx will forward the requests to the Laravel Octane service.

The nginx configuration for PHP files looks as follows:

```
location ~ \.php$ {
    include snippets/fastcgi-php.conf;
    add_header X-Debug-Config2 FASTCGI;
    fastcgi_pass unix:/run/php/php8.1-fpm.sock;
}
```

This typical configuration with nginx for using the PHP uses the FPM module via FastCGI. As mentioned in *Chapter 1, Understanding the Laravel Web Application Architecture*, nginx is not able to run PHP scripts. To do that, nginx can be configured to forward the requests to the FPM module. The FPM module will run the PHP scripts and will send back the response to nginx. All the requests to PHP files are forwarded to the FPM socket communication channel (fastcgi_pass).

In case you use Laravel Octane, you don't need to use the FastCGI option to execute PHP scripts because the PHP scripts will be executed by the Laravel Octane service via the application servers (Swoole or RoadRunner).

If you have Laravel Octane, you can start the Octane service via the `php artisan octane:start` command and then instruct nginx to act as a proxy, forwarding all the PHP requests to the Laravel Octane service.

To set nginx as a proxy, you can set this configuration in your nginx configuration file:

```
location /index.php {
    try_files /not_exists @octane;
}
location / {
    try_files $uri $uri/ @octane;
}
location @octane {
        add_header X-Debug-Config2 POWEREDBYOCTANE;
        set $suffix "";

        if ($uri = /index.php) {
            set $suffix ?$query_string;
        }

        proxy_http_version 1.1;
        proxy_set_header Host $http_host;
        proxy_set_header Scheme $scheme;
        proxy_set_header SERVER_PORT $server_port;
        proxy_set_header REMOTE_ADDR $remcte_addr;
        proxy_set_header X-Forwarded-For
                        $proxy_add_x_forwarded_for;
        proxy_set_header Upgrade $http_upgrade;
        proxy_pass http://127.0.0.1:8000$suffix;
}
```

The `try_files` line is very important for splitting the behavior of nginx regarding how to manage static resources (existing files) and dynamic resources (not existing files).

If the request is for `/test.png` and the file exists in a public directory (because of the root directive), the file is loaded and served by nginx.

If the request is for /test, it falls in the latest scenario of try_files labeled as @octane.

Then, there is a specific section to manage all requests labeled as @octane. All these requests are sent to Laravel Octane via the proxy_pass directive.

As you can see, all the FastCGI directives are commented out, or they are removed. Now, we are using the proxy_pass directive. The proxy_pass directive refers to the IP address and the port where the Laravel Octane service is running.

To make the proxy configuration work, you have to set some additional parameters such as proxy_http_version 1.1; for setting the protocol version and other options for controlling the headers.

Once you edit your nginx configuration file, you must restart the nginx process to apply your changes. Before restarting nginx, I suggest checking the syntax of the changed configuration files.

Using the nginx -t command shows you whether there are any issues with your new configuration:

```
# nginx -t
nginx: the configuration file /etc/nginx/nginx.conf syntax is
ok
nginx: configuration file /etc/nginx/nginx.conf test is
successful
```

Once you have a successful message, you can safely restart nginx. How to restart the service depends on your distribution. In the case of Ubuntu, you can use the service command:

```
service nginx restart
```

In general, if your Linux distribution doesn't provide a service command, you can use the systemctl command:

```
sudo systemctl restart nginx
```

The sudo command is used because you need to have root permission.

Once you reload nginx, you have to launch the artisan command for launching Laravel Octane.

Launching Octane

You can manually execute the octane:start command for a quick configuration test. Then, we will see how to launch Octane automatically when the system starts.

For starting Octane, you should use the octane:start command:

```
php artisan octane:start --server=swoole
```

This command starts Octane and accepts connections on localhost.

Suppose you need to execute Octane on a different machine (one machine used for running nginx, and another machine used for running Laravel Octane): you should allow Octane to accept all incoming requests via the --host parameter:

```
php artisan octane:start -host=0.0.0.0 -server=swoole
```

The 0.0.0.0 address means Octane will accept all incoming requests from all IP addresses (not just the local ones).

Launching Octane via Supervisor

If you need to launch and monitor a process during the boot of the system, I suggest doing it with Supervisor software.

There is excellent open source software named Supervisor that you can install via your GNU/Linux package manager through a command such as the following:

```
apt-get install supervisor
```

Once Supervisor is installed, you can set the startup script file for Laravel Octane.

> Supervisor
>
> If you are interested in Supervisor for other projects/commands (not just for Laravel Octane), you can retrieve more information on the official website: http://supervisord.org/.

The startup scripts managed by the supervisor are stored in the conf.d directory, which is inside the /etc/supervisor/ directory.

Create a new file, /etc/supervisor/conf.d/laravel-octane.conf, with this configuration:

```
[program:laravel-octane]
process_name=%(program_name)s_%(process_num)02d
command=php /var/www/mydomain/htdocs/artisan octane:sta-
--server=swoo- --max-requests=10- --workers- --task-workers=-
--port=8000
autostart=true
autorestart=true
user=deploy
redirect_stderr=true
```

```
stdout_logfile=/var/www/mydomain/htdocs/storage/logs/laravel-
octane-worker.log
```

Here, you have to set some directives:

- `process_name`: This is the name used for tagging the process
- `command`: This is the command line for launching the Octane service with `octane:start`
- `autostart` and `autorestart`: This is to set how and when to start the process (automatically)

  The user used for launching the process (I'm using a specific one such as `deploy`) is usually `www-data`, which, by convention, is the user used for running the web server process (nginx or Apache).

- `redirect_stderr`: If the error messages have to be redirected to the log this is used.
- `stdout_logfile`: This is the file name used for the log. Make sure that one row for each request managed by Octane is created by default.

Supervisor supports many parameters if you are interested in a more complex configuration. All the parameters are documented here: `http://supervisord.org/configuration.html`.

Once you have set the configuration file, you can update the Supervisor configuration, adding the Laravel Octane configuration:

`supervisorctl update`

Then, you restart Supervisor:

`supervisorctl restart all`

Or you can specifically restart just the Octane service:

`supervisorctl restart laravel-octane:laravel-octane_00`

This way, the Laravel Octane is launched according to the parameter defined on the command line (the command directive in the Supervisor configuration), and Laravel Octane is relaunched in case of reboot or crash.

Now, Laravel Octane has launched automatically. If you change your application's source code or deploy a new version of your Laravel application, you have to reload the worker so that all the changes are applied to all the workers.

Reloading the workers

Every time you make some changes to your application logic, you have to restart the application with the following command:

```
php artisan octane:reload
```

The `octane:reload` command restarts the workers instanced by Octane, and in this way, the code is reloaded.

```
deploy@vultr:/var/www/mydomain/htdocs$ php artisan octane:reload

    INFO   Reloading workers...
deploy@vultr:/var/www/mydomain/htdocs$
```

Figure 7.2: Reloading the workers with php artisan octane:reload

If you want to know about the status (running or not) of the Laravel Octane service, you can use the `octane:status` command:

```
php artisan octane:status
```

`octane:status` shows the **Octane server is running.** message if everything is fine. Otherwise, it shows **Octane server is not running.**:

```
deploy@vultr:/var/www/mydomain/htdocs$ php artisan octane:status

    INFO   Octane server is running.
deploy@vultr:/var/www/mydomain/htdocs$
```

Figure 7.3: Checking the status of Laravel Octane

Listening events

Laravel Octane internally defines some events for notifying and managing when something happens, such as the worker starting, a request being received, and so on. In a framework, if you can produce some events, as a developer, you need a mechanism to listen for when an event occurs. The Laravel framework provides the developer with a mechanism to define and create events and a mechanism to listen (a listener). The listener in Laravel can execute a function defined by the developer.

So, in this case, using the event listener mechanism provided by Laravel, you can listen to some specific Laravel Octane events.

Looking into the Laravel Octane config file (`config/octane.php`) in the `listener` section, you will see that the defined events are as follows: `WorkerStarting`, `RequestReceived`, `RequestHandled`, `RequestTerminated`, `TaskReceived`, `TaskTerminated`, `TickReceived`, `TickTerminated`, `OperationTerminated`, `WorkerErrorOccurred`, and `WorkerStopping`.

For example, if you want to implement a listener for the `RequestReceived` event, you can use the `make:listener` command to create the listener class:

```
php artisan make:listener RequestReceivedNotification
--event=RequestReceived
```

The `make:listener` command creates the `RequestReceivedNotification` class file based on the `RequestReceived` event.

The file is created in `app/Listeners/RequestReceivedNotification.php` with the constructor and the `handle()` method, so you can add your logic to track some of the activity in the `handle()` method; for example, I'm going to log the IP address where the request comes from:

```php
<?php

namespace App\Listeners;

use Illuminate\Support\Facades\Log;
use Laravel\Octane\Events\RequestReceived;

class RequestReceivedNotification
{
    /**
     * Create the event listener.
     *
     * @return void
     */
    public function __construct()
    {
        //
    }

    /**
     * Handle the event.
```

```
 *
 * @param   \Laravel\Octane\Events\RequestReceived;
    $event
 * @return void
 */
public function handle(RequestReceived $event)
{
    Log::info('Request Received by
            '.$event->request->ip());
}
}
```

To enable the listener, you have to add the class to the `config/octane.php` file:

```
RequestReceived::class => [
    ...Octane::prepareApplicationForNextOperation(),
    ...Octane::prepareApplicationForNextRequest(),
// Add the RequestReceivedNotification class
    RequestReceivedNotification::class
    //
],
```

In the same way, you can also add other events you want to track.

Now that we have configured the web server for the production environment, we can walk through the Laravel configuration for the production environment.

Prepping a Laravel application for production

Now, we will focus on Laravel-specific configurations for the production environment. We will see how to install production PHP/Laravel packages, avoiding installing packages for development and debugging. We will see how to optimize cache configurations, routing settings, and view optimization. Finally, we will see how to inhibit debugging options. Debugging options are very useful in development, but can slow down the execution of the Laravel application in production.

Installing the packages for production

In the `composer.json` file, two types of packages are listed: `require` defines the packages for running the application in a production environment, and `require-dev` defines the packages for running the application in a development environment.

By default, `composer` installs all the packages from both lists.

If you only want to install the packages from the `require` list (production), you can use the `--no-dev` option while you are installing the packages for the production environment:

```
composer install --optimize-autoloader --no-dev
```

Other than the `--no-dev` option, we are using `optimize-autoloader`. The option generates a data structure (an index) to map the class name, with the name of the file to be loaded.

The `optimize-autoloader` option reduces the bootstrap time.

Caching config, routing, and views

To load some critical components such as the configuration files, the routing configuration, and the view blade templates quickly, you can use the cache commands:

```
php artisan config:cache
php artisan route:cache
php artisan view:cache
```

You have to execute these commands every time you change certain files in your Laravel application in the production environment.

```
deploy@vultr:/var/www/mydomain/htdocs$ php artisan config:cache

   INFO   Configuration cached successfully.

deploy@vultr:/var/www/mydomain/htdocs$ php artisan route:cache

   INFO   Routes cached successfully.

deploy@vultr:/var/www/mydomain/htdocs$ php artisan view:cache

   INFO   Blade templates cached successfully.
```

Figure 7.4: Caching configs, routes, and views

Disabling Debug mode

When developing the Laravel application, the configuration is typically set to Debug mode. The parameter that controls the Debug mode is stored in the .env file, and the parameter is as follows:

```
APP_DEBUG=true
```

To check the status of your application, you can use the following command:

```
php artisan about
```

The about command shows the information about **Debug Mode**:

```
deploy@vultr:/var/www/mydomain/htdocs$ php artisan about

  Environment
  Application Name ........................................................... Laravel
  Laravel Version ............................................................ 9.31.0
  PHP Version ................................................................. 8.1.2
  Composer Version ........................................................... 2.4.2
  Environment ................................................................. local
  Debug Mode ................................................................ ENABLED
  URL ..................................................................... localhost
  Maintenance Mode .............................................................. OFF
```

Figure 7.5: The about command for showing the Debug mode status

To change the setting, you can change the value of the APP_DEBUG parameter to false:

```
APP_DEBUG=false
```

The configuration of APP_DEBUG set to false allows you to skip all the configurations in the code for debugging purposes (collecting and showing detailed information).

Because you changed the configuration and the configuration is cached, the suggestion is to clear the configuration to see the changes in the application:

```
php artisan config:clear
```

You should also reduce the number of log messages. Log messages are very useful for debugging purposes, but tracking the logs could be an expensive operation and could slow down the application.

For example, in the .env configuration file, typically, you can find the following:

```
LOG_LEVEL=debug
```

The levels of debugging, from most detailed (with more logs) to just tracking the errors, are as follows: `debug`, `info`, `notice`, `warning`, `error`, `critical`, `alert`, and `emergency`. My suggestion is to change the level of logs (`LOG_LEVEL`) in a production environment to `error` in order to only track critical information because reducing the number of log messages makes your application faster and can reduce the amount of storage needed.

The general suggestion is to have a specific configuration for the production environment, so the advice is to create a file, `.env.prod`, where you can store your production configuration. Then, copy the `.env.prod` file on each deployment to your `.env` file in the production environment.

Deployment approaches

You can choose a deployment strategy from the numerous ones that exist.

The rule that I would recommend when choosing a deployment strategy is to avoid installing valuable tools for the build phase in the production environment. Because the approach is to limit the responsibility of the production environment to the delivery of ready-to-deploy assets, responsibility for executing the build process should not be the responsibility of the production environment. That's why it's possible to avoid installing build tools in the production environment.

For example, I suggest avoiding performing the build of JavaScript and CSS, running tests or "linting" processes, or performing static code analysis in production environments.

In other words, it is necessary to transfer optimized code and ready-made configurations directly to the production server.

This means that if there should be build operations that aim to prepare files and configurations for production, these should be performed in a dedicated environment or a CI/CD runner.

The build mechanism and file transfer commands in the production environment can be complex. Therefore, the advice is to formalize this list of commands in a scripting tool.

I use `Makefile` because this allows me to define the individual steps to be executed but, more importantly, determine the dependencies between steps.

Typically, I define a step for compiling assets, a step for testing, a step for running `phpcs` (lint), a step for performing static code analysis, and a step for transferring files to the production environment. There are also specific steps for executing commands directly on the production server should the need arise. This number of granular steps allows me to keep each step simple.

There are several methodologies for copying and transferring files to production; the advice is to find the one that is most suited to you; typically, I use **Secure Shell (SSH)** protocol and tools.

Nevertheless, again, the most crucial part is to understand that, irrespective of the way you transfer the files or the tool you are using, you should transfer files and configuration ready to be served by the production environment.

Makefile

`Makefile` is a file used by the `make` tools to execute commands. The format is quite simple; each group of commands is grouped into steps. You can specify the step to be executed from the command line when you start running the file via the `make` command.

In the `Makefile`, all the commands are listed, and the command to be executed needs a parameter such as `SSH_USER` or `WEB_DIR`. My suggestion is to define the parameters in an external `Makefile`. Then, you can ignore (in the Git repository) the `Makefile` used for the parameters and commit the `Makefile` with the steps.

In the root directory of your Laravel project, create a file named `Makefile.param`:

```
SSH_ALIAS=some.vm
SSH_USER=deploy
WEB_DIR=/var/www/mydomain/htdocs
WEB_USER=www-data
WEB_GROUP=www-data
# DRY_RUN=--dry-run
DRY_RUN=
```

The meaning of the parameters is quite self-explanatory; you can define the SSH alias and the SSH user to access the remote machine via SSH, and then you can specify the remote directory (on the server) where your Laravel application is stored (`WEB_DIR`). Then, you can set the user (`WEB_USER` and `WEB_GROUP`) on the server, used for running the processes on the server. Then, with `DRY_RUN`, you can see (and copy) the files to be copied on the server.

For copying the file on the server, we will use `rsync`. `rsync` is a great tool for only copying the changed file.

To create a `Makefile`, create a file named `Makefile.deploy` and fill it with the following:

```
.PHONY: help all remotedu rsynca copyenvprod fixgroupuser
newdeploy migratestatus migrate migrateseed migraterefresh
buildfrontend optimize composerinstallnodev restartworkers
octanestatus installdevdeps deploy


include Makefile.param.prod


all: help


help:
    @grep -E '^[a-zA-Z_-]+:.*?## .*$$' $(MAKEFILE_LIST) |
```

```
        sort | awk 'BEGIN {FS = ":.*?## "};
        {printf "\033[36m%-30s\033[0m %s\n", $$1, $$2}'

remotedu: ## Execute DU command in htdocs dir, just for
diagnostic purpose
        ssh ${SSH_ALIAS} "cd ${WEB_DIR}; du -h"

rsynca: ## execute Rsync from current dir and remote htdocs
${WEB_DIR}
        rsync ${DRY_RUN} -rlcgoDvzi -e "ssh" --delete .
          ${SSH_ALIAS}:${WEB_DIR}   --exclude-from
          'exclude-list.txt'

copyenvprod:
        scp .env.prod ${SSH_ALIAS}:${WEB_DIR}/.env

fixgroupuser: ## Add the right group(www) to the deploy user
(ssh user)
        ssh -t ${SSH_ALIAS} "sudo usermod -a -G ${WEB_GROUP}
        ${SSH_USER}"

fixownership: ## fix the ownership for user ${WEB_USER} into
${WEB_DIR}/storage
        ssh -t ${SSH_ALIAS} "sudo chown -R
        ${WEB_USER}:${WEB_GROUP} ${WEB_DIR}/storage; ls -lao
        ${WEB_DIR}/storage"

newdeploy: buildfrontend rsynca copyenvprod fixgroupuser
fixownership migrate ##first deploy

migratestatus: ## list the migration status
        ssh ${SSH_ALIAS} "cd ${WEB_DIR}; php artisan
        migrate:status --env=prod"

migrate: ## Execute migrate command for DB schema
        ssh ${SSH_ALIAS} "cd ${WEB_DIR}; php artisan migrate
        --env=prod"
```

```
migrateseed: ## Execute migrate command for DB schema
    ssh ${SSH_ALIAS} "cd ${WEB_DIR}; php artisan
    migrate:refresh --seed --env=prod"

migraterefresh: ## Execute migrate command for DB schema
    ssh ${SSH_ALIAS} "cd ${WEB_DIR}; php artisan
    migrate:refresh"

buildfrontend: ## execute npm task to compile frontend assets
(js and css...)
    npm run build

optimize: ## Optimize application in production
    ssh ${SSH_ALIAS} "cd ${WEB_DIR}; php artisan
    config:cache; php artisan route:cache; php artisan
    view:cache"

composerinstallnodev:
    composer install --optimize-autoloader  --no-dev

restartworkers:
    ssh ${SSH_ALIAS} "cd ${WEB_DIR}; php artisan octane:reload"

octanestatus:
    ssh ${SSH_ALIAS} "cd ${WEB_DIR}; php artisan octane:status"

installdevdeps:
    composer install
    npm run dev
    php artisan config:clear


deploy: buildfrontend composerinstallnodev rsynca copyenvprod
migrate restartworkers installdevdeps
```

The last line is for the `deploy` step. The `deploy` step doesn't include commands to execute but consists of a list of steps to invoke sequentially.

The first step executed will be `buildfrontend`. This step includes the `npm run build` command. The command will build the frontend assets.

The next step included in the `deploy` step is `composerinstallnodev`. The command executed by `composerinstallnodev` will install the packages for the production (skipping the dev packages):

```
composer install --optimize-autoloader   --no-dev
```

Then, there is the step for copying the files on the server: `rsynca`. The step includes a command:

```
rsync ${DRY_RUN} -rlcgovzi -e "ssh" --delete . ${SSH_
ALIAS}:${WEB_DIR}  --exclude-from 'exclude-list.txt'
```

In the command, some variables (replaced with the values loaded from `Makefile.param.prod` file) are used.

The `Makefile.param.prod` file is loaded at the beginning of the `Makefile` thanks to the directive including the following:

```
include Makefile.param.prod
```

The `rsync` command uses a list of options:

- `-r` : Recursive into directories.

- `-l`: Copies symlinks as symlinks.

- `-c`: Copies only the changed files. The changes are detected with the `-c` option by comparing the checksum of the source and the target file.

- `-g`: Preserves the group during the copy (owner).

- `-o`: Preserves the user during the copy (owner).

- `-v`: Verbose output.

- `-z`: Compresses files during network transfer.

- `-i`: Displays a summary.

The `--exclude-from 'exclude-list.txt'` option excludes from the copy all the files listed in the `exclude-list.txt` file.

The exclude file contains the following:

```
.git
.sass-cache/
Makefile*
exclude-list.txt
nohup.out
composer.phar
node_modules
.env*
.idea
.editorconfig
.babelrc
.gitattributes

logs
*.log
npm-debug.log*

node_modules

storage/framework/cache/*
storage/framework/sessions/*
storage/framework/views/*
storage/logs/*
storage/app/*
storage/testing/*

exports/*
app/config/prod/*
app/config/stage/*
resources/assets
public/files
public/chunks
public/storage
```

```
bootstrap/cache/config.php
```

```
.phpunit.result.cache
```

The `exclude-list.txt` file lists all the files and directories that are not useful in the production environment. Typically, the excluded files/directories are package directories, storage directories, cache files, and so on.

Feel free to add the files in your application that you want to exclude and don't want to add to the production environment.

The `copyenvprod` step will copy the local `.env.prod` (where you can store your specific configuration for production) to the remote `.env`:

```
scp .env.prod ${SSH_ALIAS}:${WEB_DIR}/.env
```

The `scp` command allows you to copy files via SSH protocol.

Remember to include the `.env.prod` file in the `.gitignore` file (to exclude it from the `git` commit).

The `migrate` step executes the migrations on the remote server (thanks to the SSH command):

```
ssh ${SSH_ALIAS} "cd ${WEB_DIR}; php artisan migrate
--env=prod"
```

Then, the `optimize` step is executed. The `optimize` step will launch the caching Laravel command for routing, views, and configuration on the target machine:

```
ssh ${SSH_ALIAS} "cd ${WEB_DIR}; php artisan config:cache; php
artisan route:cache; php artisan view:cache"
```

Then, if you remember, every time you change the files in the production environment, you should restart the workers to apply the changes to the up-and-running Laravel Octane application, so the `restartworkers` step will launch:

```
ssh ${SSH_ALIAS} "cd ${WEB_DIR}; php artisan octane:reload"
```

Executing the `Makefile` from the local development environment is not suggested. Typically, you can launch the `Makefile` from the CI/CD pipeline. We execute the `Makefile` from a dedicated environment (CI/CD) for separation of concerns; the local environment is for development, the CI/CD environment is for building and executing code quality tools programmatically, and the production environment is for delivering assets and the application.

Anyway, if you are not using a CI/CD pipeline and you are launching the `Makefile` from the local copy of the sources, you must restore the dev package (for production, deploy the `composerinstallnodev` step, which will only install production packages). For restoring the dev packages, the `installdevdeps` step exists, and it will execute a standard `composer install` command and then clear the config:

```
composer install
php artisan config:clear
```

As you can see, the configuration of `Makefile` is straightforward. Because `Makefile` has strict rules about the indentation (the indentation is used to understand where a step starts and where it ends), the only recommendation is that the line indentation is done via the tab character. Once `Makefile` is ready, you can execute it.

To run the `Makefile` named `Makefile.deploy`, you can run the following:

```
make -f Makefile.deploy deploy
```

The `make` command accepts the name of the step to be executed as input. Optionally, you can also specify the `Makefile` to be parsed via the `-f` parameter.

With the `make deploy` command, the application is fully deployed on the server. The `Makefile` also includes some commands for the fine-tuning of the environment, such as fixing permissions of some directories, clearing the cache, and so on.

Summary

With this book, we explored some ways to create a reliable and fast Laravel application. We used features provided by Laravel Octane with the integration with Swoole and RoadRunner.

We involved some special features of Swoole, especially for caching, concurrent task execution, and scheduled task execution.

We explored some practical examples, and we approached the design of the application differently, delegating tasks to the external executor (via the queue).

We progressively improved the performance of the application by applying multiple techniques.

In this chapter, we explained the architecture, the configuration of the tools used in production (nginx), and the implementation of the scripts to deploy a Laravel Octane application into a production environment.

Hope you enjoyed reading this book.

Index

`Packt.com`

Subscribe to our online digital library for full access to over 7,000 books and videos, as well as industry leading tools to help you plan your personal development and advance your career. For more information, please visit our website.

Why subscribe?

- Spend less time learning and more time coding with practical eBooks and Videos from over 4,000 industry professionals

- Improve your learning with Skill Plans built especially for you

- Get a free eBook or video every month

- Fully searchable for easy access to vital information

- Copy and paste, print, and bookmark content

Did you know that Packt offers eBook versions of every book published, with PDF and ePub files available? You can upgrade to the eBook version at `packt.com` and as a print book customer, you are entitled to a discount on the eBook copy. Get in touch with us at `customercare@packtpub.com` for more details.

At `www.packt.com`, you can also read a collection of free technical articles, sign up for a range of free newsletters, and receive exclusive discounts and offers on Packt books and eBooks.

Other Books You May Enjoy

If you enjoyed this book, you may be interested in these other books by Packt:

Clean Code in PHP

Carsten Windler, Alexandre Daubois

ISBN: 978-1-80461-387-0

- Build a solid foundation in clean coding to craft human-readable code
- Understand metrics to determine the quality of your code
- Get to grips with the basics of automated tests
- Implement continuous integration for your PHP applications
- Get an overview of software design patterns to help you write reusable code
- Gain an understanding of coding guidelines and practices for working in teams

Test-Driven Development with PHP 8

Rainier Sarabia

ISBN: 978-1-80323-075-7

- Learn where and how to start a software project
- Learn how to use Jira as a tool to organize what needs to be done
- Learn when and how to write Unit, Integration, and Functional tests Using PHPUnit
- Write behavior-driven tests using Behat
- Apply SOLID principles to help write more testable code
- Learn how to get the most out of your automated tests by using Continuous Integration
- Use Continuous Delivery to help you prepare your application for deployment

Packt is searching for authors like you

If you're interested in becoming an author for Packt, please visit `authors.packtpub.com` and apply today. We have worked with thousands of developers and tech professionals, just like you, to help them share their insight with the global tech community. You can make a general application, apply for a specific hot topic that we are recruiting an author for, or submit your own idea.

Hi!

I am Roberto Butti, author of *High Performance with Laravel Octane*. I really hope you enjoyed reading this book and found it useful for increasing your productivity and efficiency.

It would really help me (and other potential readers!) if you could leave a review on Amazon sharing your thoughts on this book.

Go to the link below or scan the QR code to leave your review:

`https://packt.link/r/1801819408`

Your review will help us to understand what's worked well in this book, and what could be improved upon for future editions, so it really is appreciated.

Best wishes,

Roberto Butti

Download a free PDF copy of this book

Thanks for purchasing this book!

Do you like to read on the go but are unable to carry your print books everywhere? Is your eBook purchase not compatible with the device of your choice?

Don't worry, now with every Packt book you get a DRM-free PDF version of that book at no cost.

Read anywhere, any place, on any device. Search, copy, and paste code from your favorite technical books directly into your application.

The perks don't stop there, you can get exclusive access to discounts, newsletters, and great free content in your inbox daily

Follow these simple steps to get the benefits:

1.   Scan the QR code or visit the link below

https://packt.link/free-ebook/9781801819404

2.   Submit your proof of purchase

3.   That's it! We'll send your free PDF and other benefits to your email directly

www.ingramcontent.com/pod-product-compliance
Lightning Source LLC
Chambersburg PA
CBHW060600060326
40690CB00017B/3775